Beads
in Fashion
1900-2000

Leslie Piña,
Lorita Winfield, &
Constance Korosec

Schiffer Publishing Ltd

4880 Lower Valley Road, Atglen, PA 19310 USA

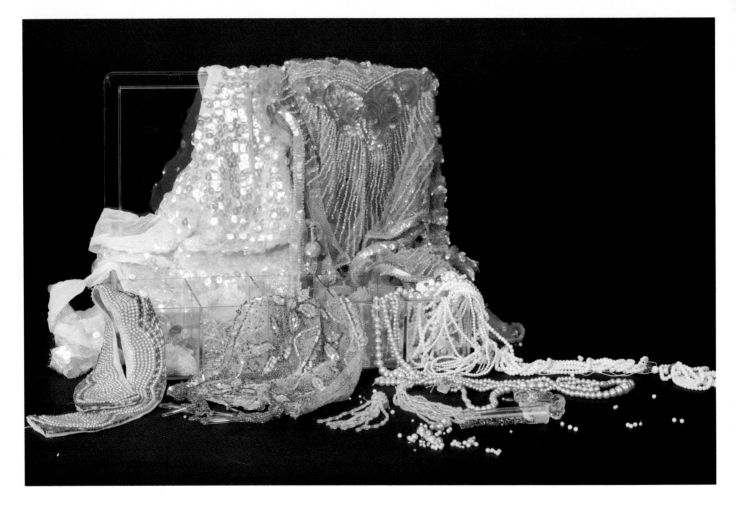

Photography by Leslie & Ramón Piña
Designed by Leslie Piña
Book Layout by Bonnie M. Hensley
Type set in Bodoni Bk BT/GoudyOISt BT

ISBN: 0-7643-0792-4
Printed in China
1 2 3 4

Published by Schiffer Publishing Ltd.
4880 Lower Valley Road
Atglen, PA 19310
Phone: (610) 593-1777; Fax: (610) 593-2002
E-mail: Schifferbk@aol.com
Please visit our web site catalog at **www.schifferbooks.com**

In Europe, Schiffer books are distributed by Bushwood Books
6 Marksbury Avenue Kew Gardens
Surrey TW9 4JF England
Phone: 44 (0)181 392-8585; Fax: 44 (0)181 392-9876
E-mail: Bushwd@aol.com

This book may be purchased from the publisher.
Include $3.95 for shipping. Please try your bookstore first.
We are interested in hearing from authors with book ideas on related subjects.
You may write for a free printed catalog.

Dedication

Dedicated to our mothers who introduced us to beads and fashion.

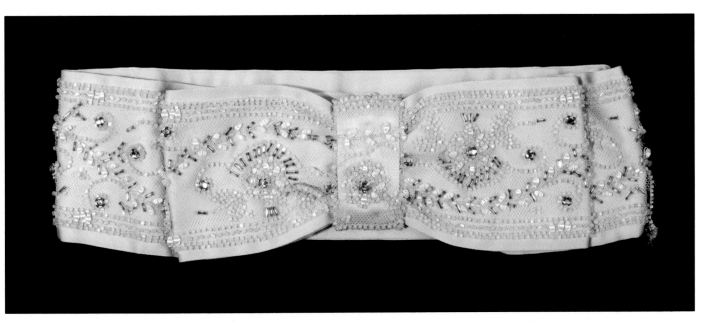

Blue satin cummerbund, hand made, from a 1961 high school prom gown. The cummerbund is layered with antique beading from the early 20th century. Lorita's mother found the lace in an antique shop and used it to decorate accessories and shoes for the formal. This began a passion for beads.

Acknowledgments

We would like to express our thanks to all anonymous sources and to the following who so generously provided beads in fashion for us to photograph for this book: Travis Abbott, Diane C. Andrica, Houri Askari, Christine Attenson and Mitchell S. Attenson of Attenson's Antiques and Books, Virginia Mallett Azzolina, Barbara Primeau Bachman, Saiede Tajbakhsh Baghery, Judith Letteer Baidy, Virginia Barry, Cynthia Barta of Legacy Antiques and Vintage Clothing and Suite Lorain Antiques, Paulette Batt of Larchmere Antiques & Collectibles and Paulette's Antiques & Vintage Clothing, Zsuzsa Csepanyi Bawab, Lillian Birnbaum Horvat for Estate of Marisa Birnbaum, Betty Weber Bolwell, Bonnie's Goubaud, Linda Bowman of Legacy Antiques and Vintage Clothing, Anne Katherine Brown of Friends Antiques, Mary Broyles of Ornamental Resources, Krystyna Bryjak, Carole Carr, Elizabeth Crowe, Jerri Lewis Dennis, Jill Korosec Dennis, Frances Dickenson, Ethel Dindia, Lisa C. Egizii, Susan Lehman Ellick, Lucia Zavarella Fedeli, Virginia Cangelosi Folisi, Shirley Friedland, Barbara Szabo Galbos, Nancy C. Goldberg, Zillah Solonche Green, Anna Greenfield, Mari Stanek Hageman, Robin Herrington-Bowen, Irene Hoffman, Lillian Birnbaum Horvat, Lauren Humphrey, Ann D. Jackson, Donna Kaminsky, Linda Katz, Giuliana C. Koch, Alice Kuo, Kuo Designs, Ruth G. Kyman for Estate of Zelta Schulist Glick, Pamela F. LaMantia, Doreen Leaf-Hund, Emma Lincoln, Gloria Azzolina Lorenzo, Cheryl Byerley Manser, Marc Goodman Antiques, Janet King Mednik, The Museum of The Fashion Institute of Technology, Jesse Oates of Nordstrom, Paula Ockner, Katherine O'Neill, Sandy Osborn, Cindy Pressler, Rock and Roll Hall of Fame & Museum, Deborah Johnson Rogers, Marilyn Ruckman, Sarah N. Sato, Arlene Schreiber, Phyllis Seltzer, Rebecca Smith, Sally Smith, Catherine Elizabeth Szabo, Margaret Thorpe, Veronica Trainer, Charlotte Michell Trenkamp, Wilma Simon Trenkamp, Russell Trusso, Ursuline College Historic Costume Study Collection, Virginia Marti Veith of Virginia Marti Fabrics, and Pamela Nickel Wurster.

Thanks again to Ramón for his assistance with the photography, to Peter and Nancy Schiffer, Jennifer Lindbeck, Bonnie Hensley, and the gang at Schiffer publishing.

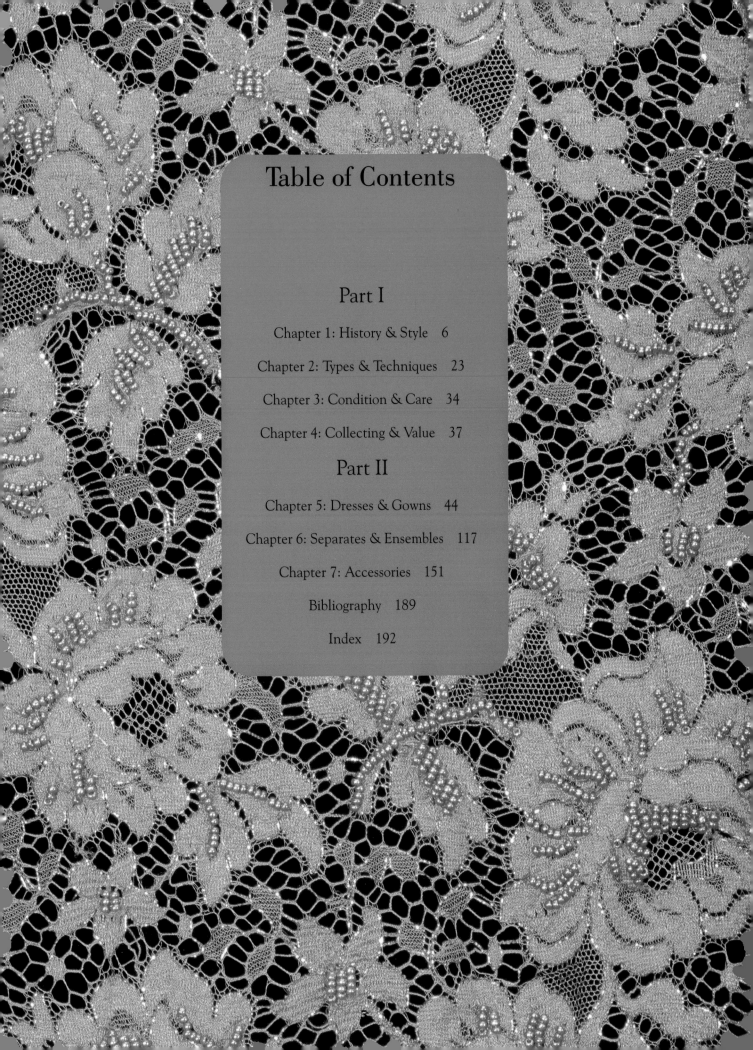

Table of Contents

Part I

Part II

PART I

Chapter 1
History & Style

Where Are Beads Made?

Glass beads: Italy, Austria, Czechoslovakia, France, Japan, India

The tiny glass beads that shimmer on the following pages are made in one or more of a number of countries that have long histories of producing glass beads, especially Italy. The island of Murano, a brief boat ride off the main area of Venice, has been dominant in the world's glass bead market since the beginning of the 15th century.

Austria, home of the Swarovski glass factory, and Czechoslovakia, or Bohemia as it was known, have legacies of craftsmen skilled at cutting and grinding facets on crystal beads.

France has been an important producer of glass beads for centuries and continues to supply the world with seed beads.

Japan, in recent years, has developed the technique to produce "uniform" glass seed beads that "fit" perfectly against one another whether woven, embroidered, or strung. The variety of colors produced are unequaled, and these beads have inspired the world's bead artists and embroidery artists.

India, since its independence in 1947, has emerged

as a major exporter of glass beads and as one of the world's most important centers of bead embroidery work.

Pearls, Jet, and Coral beads: England, Italy, the Pacific Region

The natural materials of pearl, jet, and coral have been worked into beads since pre-historic times and have long been used in embroidery and the embellishment of apparel. Their influence in the history of fashion is such that a color is associated with each of them. The word *jet* came into the English language from the French word *jais*, a generic term for beads of all kinds and colors. Reference to bead embroidery as *la broderie en jais* has been found in a French manual on the subject published in the 18th century. Translations of the word *bead* also demonstrate the importance of these materials in history: in French, *perle*; in German, *glasperle*; and in Italian, *perla*. The Dutch were particularly fond of coral beads in their early history and have as their word for bead, *kraal*.

Millefiori beads, made in Italy, c.1950s-1970s. *Courtesy Shirley Friedland*

Jet: Black beads from the wood-fossil known as jet had been popularly used on apparel for two centuries before becoming high fashion during the Victorian Era. There are several sites around the world where jet deposits are found (Spain, France, Russia, China, and the United States, in California), but the center of the jet mining and ornament industry was located in North Yorkshire, England. By 1920, the fashion for jet and the skill of carving jet with up to one hundred facets had ended.

Right: Collar of faceted jet bugle beads, hand embroidered on silk braiding, c. late 1800s.

Far right: Red silk vest with navy lace appliqué hand beaded with faceted jet bugles, c. late 1800s. *Courtesy Ursuline College Historic Costume Study Collection*

Collection of black jet, faux jet, or black glass beads and beadwork.

Coral: The calcium-like skeleton of a sea polyp is "fished" from reefs in the Mediterranean and especially in the Pacific around Japan. Coral takes hundreds of years to form. Because it is highly valued, it has been aggressively collected from the reefs around the world and will no doubt soon be in short supply. Pacific coral is larger than the coral of the Mediterranean Sea and can, therefore, be worked into larger beads. Coral work from Tibet, Nepal, and China is found in elaborate headdresses.

Like jet work, making coral beads is labor intensive and requires great skill. Coral does not harden completely until it is exposed to the air, and, at one time, the craftsmen had to work with it in water. Today, it is soaked in acid to remove the calcified outer layer, and then cut, pierced, and shaped by hand.

From medieval times coral was used in Europe mainly for the decoration of religious articles. At the end of the 19th century, coral was highly fashionable to accent bonnets. Examples of apparel embroidered with coral in this century can be found in museums.

Branch coral in various shades of pink and red.

Bottom left: Coral beads, round and rice shaped, in various shades of pink and red.

Bottom right: Beaded beads. Wooden beads covered in hand woven antique coral beads made from the tips of branch coral. *Beadwork by Alice Kuo, Kuo Designs, Los Angeles, California*

Pearls: Natural pearls continue to be prized among jewels. They are found inside oyster shells and are formed by the irritation of something foreign such as a grain of sand. For centuries divers have gathered the pearl oysters from the Pacific, the South Seas, and the Persian Gulf. Freshwater pearls are just as their name suggests—pearls from mollusks at home in rivers throughout the world.

Left: Natural Thai silk ivory wedding gown hand beaded with thousands of pearls in a poinsettia and cabbage rose design. Wallentin Collection label, made in Thailand, c.1996. *Courtesy Jill Korosec Dennis*

Right: Detail of gown.

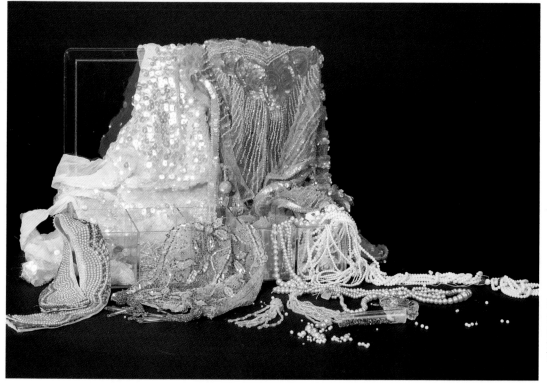

Collection of ornamental pearls, mother-of-pearl beads, sequins, embroidered collars from the 1940s, and a hand beaded net blouse, c.1920s.

From the Baroque period, pearls were in great demand in fashion and for decorating the robes of civil and religious "royalty." The imitation pearl was first made in the 15th century and finally "perfected" by the French who developed a process involving wax in the mid 17th century. These pearls, called Roman pearls, were produced until the beginning of the 20th century.

The Japanese first produced what is known as the cultured pearl in the late 19th century. A round mother-of-pearl bead is placed in the shell of the pearl oyster, and the man-made irritant is covered over by nacre to make the luster. Japan also exports an excellent quality of glass pearls to meet the current demand for pearls in beaded fashion.

The cost of beading today with jet, real pearls, or coral would be prohibitive. Examples can be found in collections and museums.

Top: Bridal tiaras hand embroidered with faux pearls and waxed orange blossoms, c.1920s-1930s. *Courtesy Anna Greenfield*

Bottom: Detail of tiaras.

Egyptian, Middle Eastern, Italian, African, American Indian

Bead making and the wearing of beads are believed to have been male endeavors until relatively recent times. Beads were associated with personal identity, spiritual power, protection, strength, and hunting prowess. They also made a man more attractive. Whoever picked up that first seed, nut, berry, or bone and looked at it as a beautiful "find" to keep and wear started a trend. Beads became the fashion!

As man discovered how to drill small holes in his beads, to make coarse thread from sinews and bark, and hooks and needles from slithers of bone, he was able to improve and perfect his technique. He learned to stitch, knot, form nets, and weave. With these skills, he made the first textiles and decorated them with shells, bones, and small stones. This was the first bead embroidery, which dates to between 10,000 and 8,000 B.C.

Egypt and the Middle East: The wealth and stability of the ancient Egyptian civilization enabled beads to be manufactured in quantity and from a variety of materials unequaled by any other civilization. Beginning in Old Kingdom Egypt, about 2200 B.C., beads were stitched to linen and papyrus and were used to decorate sandals, belts, and other apparel such as aprons and skirts. Not until recently have beads become as fashionable as in ancient Egypt–worn by the members of both sexes, and by a diverse cross section of society.

It was the development of *faience*, a ceramic made with quartz, sand, and colored glaze, that made beads accessible to all social levels in Egypt. By copying the use of gold, turquoise, and other precious and semi-precious beads worn by royalty, even ordinary Egyptians could be extravagant with this shiny man-made material.

Beads were buried with their owners, and the types and quality of beads found in tombs indicate the social status of the person and the family. *Sha* is the Egyptian word for "luck" and *sha-sha* is "bead." Perhaps this explains the Egyptian custom of wearing beads from head to toe.

Egyptian and Mid-Eastern artisans working to produce the faience necessary to satisfy this demand are the probable developers of glass. These first glass makers must have been delighted to see that glass beads more closely resembled the gems and precious stones they were attempting to imitate.

The history of ancient glass making includes three major periods: the Egyptian New Kingdom (c.1650 - 1085 B.C.), the Roman Period (c.100 B.C. to A.D. 400), and the Islamic domination of the Mediterranean (c.A.D. 600 - 1400).

Italy: Although there is evidence of glass making in Venice hundreds of years earlier, the period between 1200 and 1400 is regarded as the foundation of the glass industry. All of the glass factories were relocated to the island of Murano in 1292 to protect Venice from the risk of fires caused by the continually burning furnaces. More importantly, the glass industry was moved to keep the glass making techniques secret. The glass makers' guild was directly under the jurisdiction of the highest governing body of the Venetian Republic, the Council of Ten, and, for the next half century, Murano glass makers were governed by a very strict "non-compete" clause. Under penalty of death, they were forbidden to share the secrets of their glass making, to leave the island, or set up glass making factories elsewhere. Although glass factories existed in other European countries at the time, Venice dominated the industry and the market.

During the years of exploration, discovery, and European expansion, glass beads were produced in the furnaces of Murano, Holland, and Bohemia (Eastern Europe). They were then exported in volume and used as gifts and barters on the trade routes to Asia, the Far East, Africa, and the

Which Cultures Have Long Traditions of Making and Using Beads?

Sea shells strung as necklaces.

Turquoise colored beads.
Left: Chinese cloisonné beads, contemporary.
Center back: Egyptian faience beads, contemporary.*Courtesy C. Diane Andrica*
Center front: Glass trade beads found in Southwestern United States, c.1850s.
Right: Chinese Peking glass beads, c.1920s.

Americas until the mid 19th century. The African and the North and South American Indian cultures had histories of bead making and beadwork, but marveled at the new glass beads (trade beads) brought by the European traders. With these large and small beads, they crafted jewelry and apparel that is still admired and collected today.

Aside from adorning the robes of kings and queens, much of the embroidery using beads began with the decoration of religious articles. Made with gold, pearls, coral, and gems, this embroidery was used to decorate both the church altars and the priests. In England, during the Elizabethan period, portraits of wealthy men and women of status show them wearing large quantities of pearls and precious gems on their clothing. As in ancient Egypt, the use of glass beads, bone, and painted wood allowed the middle class to adopt the latest fashion and embellish their clothing as well.

By the 17th century, the practice of working with beads was considered an essential craft for young women. The uses for beads increased to include needle work enhanced with beads, and objects and accessories, such as purses, made entirely from beads. Influenced by the opulence of the French Court, the 18th century saw lavish fashions worn by the aristocracy, literally encrusted with gold threads and beads of all kinds. Toward the end of the 18th century, fashion be-

came a political issue when the populace was restricted by court decree to wear black wool devoid of any embellishment. After the French Revolution, all European fashions became less elaborate.

North America: For two hundred years, glass beads brought by explorers were used by the indigenous cultures of North America. As Indian glass beadwork evolved, some Indian cultures incorporated European patterns for their own tribal use.

Eventually, artifacts were designed and produced specifically for Europeans who had developed an interest in Indian beadwork. By the late 19th century, bead embroidery and apparel decorated with beads had once again become important to fashion. Indian beadwork was so popular by the beginning of the 20th century that Mary White's manual *How to Do Beadwork* begins: "The present interest in beadwork undoubtedly sprang from our enthusiasm for Indian handicrafts." It was not until the late 1970s that American Indian beadwork and glass beads in Indian motifs became well established in popular fashion.

Opposite page: Detail.

Below: American Indian beaded purses.
Top: Hand beaded on wool flannel in lazy stitch and couched bead embroidery, c. late 1800s.
Bottom: Hand beaded on velvet in lazy stitch and couched bead embroidery. This purse was made especially for the tourist trade of large seed beads, c.1930s.

19th-century Tibetan head ornament made of coral beads with turquoise stones.

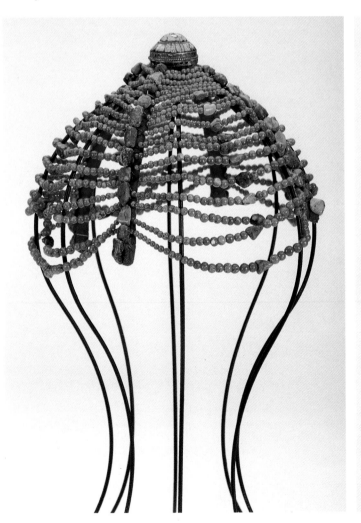

What Exactly Are Trade Beads?

Sought after by collectors, these beads come in all sizes and colors and are mainly of European glass. They were used to trade with indigenous cultures as the European exploration and exploitation of the world began in earnest in the 16th century. Great volumes of glass beads were brought to Africa throughout the 19th century. European glass factories continued to produce beads for trading in colonial Africa well into the next century. Peace Corps volunteers were among those who went to Africa in the 1960s and brought out examples of beautiful trade beads, which now are in such great demand.

In North America, trade beads were used as currency with the American Indians for the acquisition of everything from fur to land. A trail of trade beads can be found around the globe in areas marked by the old trade routes. Due to individual tribal preferences, beads found in the Americas usually differ from those found in Africa.

Trade beads were also made in India and China along many of the stops of the major trade routes. The larger beads are works of art and are usually wound: made individually by melting glass canes around a rod to create the size, color, and designs. The smallest beads were drawn, then cut from a long cane of blown glass, and sold and traded by weight. These so-called "pound" beads were eventually used to decorate fabric.

Trade beads, antiques mixed with contemporary reproductions. *Courtesy Shirley Friedland*

Alaskan Indian trade beads including dentilium shells and Russian blue beads. *Courtesy Frances Dickenson*

Are Beads Usually Found in the Place Where They Were Made?

No, and that is part of the delight in collecting and researching beads and beadwork. The origins of some beads are pure mysteries, since they may have been traded a number of times in different places around the world. The trail of one bead may, in fact, circle the world!

By studying the trade routes, beads can be traced from China, to the Philippines, to the United States. Sometimes called *padre* beads, these were ultimately distributed by the Spanish priests (padres) and found near missions.

The larger beads, though difficult to be certain of their origin, are easier to trace than the smaller seed beads found embroidered on fabric. Many times the type or style of the design made with the beads may help determine the culture, though not necessarily the origin of the beads themselves.

Can a Beaded Item Be Dated?

Even the experts in charge of collections of vintage beading admit that it is difficult to be exact in establishing the date of an item. Fortunately, there are resources aside from museums that serve as guides for dating beadwork that is from the last one hundred fifty years. During the 19th century, it was highly important for women of any age or social status to be knowledgeable about crafts, especially those designated as "feminine," such as needlework and bead embroidery. Magazines provided instructions, specific designs, and patterns for beaded items used for both personal apparel and interior decoration.

Although not very reliable as a specific dating method, the size of a bead can indicate the period of its manufacture. For example, the "pound beads" of the 18th century are larger than those of the early 19th century, a time when minute glass beads became popular. However, later that same century, beads were again produced and worked in larger sizes.

Inventions and techniques also can be helpful in dating an item. Machines for making simple embroidery stitches were used as early as 1828, and the sewing machine followed in the middle of the 19th century. Tambour beading was developed by the French toward the end of the 19th century, and at that time the first machines designed to attached beads to fabric were invented. This was largely due to the influence and the importance of the *haute couture* fashion houses in Paris. However, most dresses were still made and beaded by hand well into the 20th century. Thus, determining whether an item of clothing is hand beaded or machine beaded does not necessarily determine its date.

Style or design of an item can also be of help when dating a piece. Though, recently, styles have been re-cycled so much that dating an item according to its style has become a challenge. Even nearing the end of the 20th century, designers continue to re-introduce styles to satisfy the demand for "retro." In order to be certain, one must examine the fabric, consult fashion magazines and books, and visit museums with costume collections and exhibits.

Can Determining the Date of an Item Help Identify the Place of Origin? Can Determining the Place of Origin Help Date a Piece?

Yes is the answer to both questions. The date or place of origin of a piece provides a context and starting point from which the history of a particular beaded piece can then be determined. Further study of a piece's history can then lead to a fairly accurate description of the piece's origin and vintage.

Are Certain Designs More Suited to Beaded Ornament?

Yes. Machines that had been used for applying thread embroidery and for making designs on trims (passementeries) were adapted to work with beads by the end of the 19th century. Machine embroidery used linear patterns that could easily be transferred to beadwork. One particular design named from the Italian *vermicelli*, or worm pattern, was familiar on machine embroidered braid work and is seen in fashions from the beginning of the 20th century.

Sample books from that time show many different patterns that were possible on the Cornely machine. This was a single thread machine that worked a chain stitch and could control manually the direction of the pattern. It could be adapted for several different techniques, such as couching, and to apply beads and sequins by using a universal feed method. This machine is still used in the fashion industry today.

Detail of machine tambour beading in "vermicelli" pattern, c.1920s.

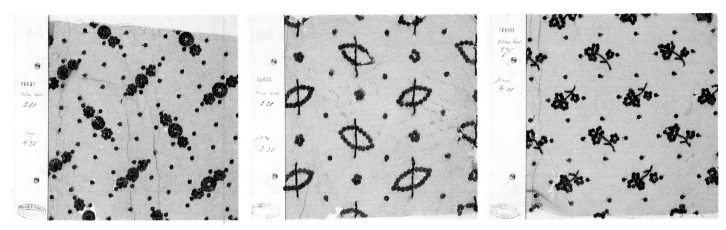

Sales samplers of hand beaded silk netting. Diersch & Schmidt, Ribonstock I. S., label, c. early 1900s.

Are There Classic or Traditional Designs in Beading That Transcend Fashion?

Even current fashion catalogs include beading designs that closely resemble the patterns used by the ancient Egyptians. For example, a pattern resembling feathers that can be seen on the clothing of Egyptian statues from c.2000 B.C. is repeated in pearls and glass beads on a flapper dress from c.1925.

Another pattern is the classic diamond-shaped design of knotted nets, which was worked in faience bugle beads by the Egyptians. This ancient pattern can be found in a 20th-century collar, a dress for a pop star, and on the hem of a formal dress.

A third characteristically Egyptian fashion is the broad beaded collar, which reappears on a 1970s evening dress.

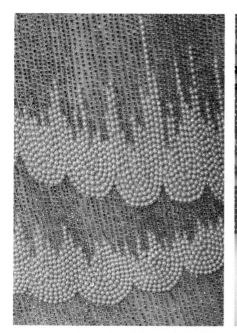 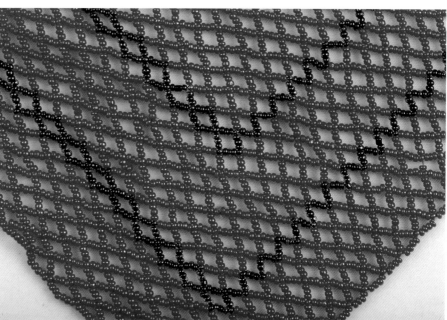

Top left: Detail of hand tambour beading in repeated "feather" pattern, c.1920s.

Top right: Detail of hand beaded, off-loom weaving in "netting" or "lattice" pattern.

Bottom left: Museum display of beaded dress belonging to Tina Turner, hand beaded, woven off-loom in "netting" or "lattice" pattern. *Courtesy Rock and Roll Hall of Fame & Museum, Cleveland, Ohio* Photo by Hal Stata

Bottom right: Bodice of evening gown influenced by Egyptian broad collar design, c.1970s. *Courtesy Ursuline College Historic Costume Study Collection*

Why So Much Black on Black in Victorian Beading?

The abundance of black clothing and accessories from the late 19th century can be explained by the death of Prince Albert of England in 1861. Queen Victoria went into mourning for the rest of her life, and the upper and middle classes in England, as well as many English subjects world wide, were required to dress in mourning attire. They even dyed clothing to keep up with the demand for black. Formal mourning attire affected the clothing industry and continued beyond the death of Victoria in 1901 and Edward VII in 1910. As late as the 1950s, travel records kept for the British royalty list complete mourning ensembles as a part of their travel luggage in case a ranking individual happened to die.

and ornamentation is understandable considering the length of time, almost two years, women were restricted from wearing fashionable colors and embellishments. In her *The Bead Embroidered Dress*, Joan Edwards quotes the editor of *The Ladies' Treasury*, a publication of 1882: "When black is worn it is loaded with jet."

The third period, "half-mourning," followed for six months to a lifetime, depending on an individual's choice. Although she could wear the fashions of the day, the widow was required to select fabrics in pale grays or mauve, for example, and use subtle decorations such as crystal beads.

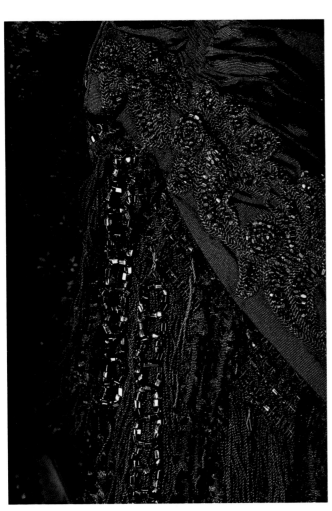

Left: Detail: hand embroidered ornamentation of jet beads on silk braiding.

Right: Mourning ensemble, hand beaded in black jet, c. late 1800s. *Courtesy Ursuline College Historic Costume Study Collection*

Periods of mourning were designated by degree. During "deep" or first mourning period, no area on the dress could have a shine, and it was mostly crepe, all black, right down to the underwear. This lasted twelve months and one day.

The second period lasted another nine months, and, although no longer required to be covered in crepe, the widow attached trims of this fabric to her dress and hat. It was during this period of mourning that the addition of jet (black) beads and trimming were allowed. The excessive use of jet beading

The jet industry of Whitby, England, flourished from 1860 to the first decade of the 20th century. Jet was carved into all sorts of objects, from fruit to hat pins, and the peak of the production is noted as from 1870-72. The demand in England, on the continent, and throughout the world soon depleted the supply of the best quality hard Whitby jet. An alternative was an inferior (soft) jet from the area or imported jet from Spain; both of which were quick to lose their polished finish and broke easily. Imitation jet satisfied the demand by the 1920s.

Top left: Velvet cape. Fur trim with hand embroidered ornamentation of jet beads on brown braided silk appliqué, c. late 1800s. *Courtesy Anna Greenfield*

Bottom left: Bodice in silk. Hand embroidered in jet beads on braided silk appliqué, c. late 1800s. *Courtesy Ursuline College Historic Costume Study Collection*

Top right: Detail of velvet cape.

Bottom right: Bodice in silk. Hand embroidered in jet beads on braided silk appliqué, c. late 1800s. *Courtesy Ursuline College Historic Costume Study Collection*

Top left: Velvet cape. Fur trim with hand embroidery in jet beads on braided silk appliqué. James McCreery & Co., New York label, c. late 1800s. *Courtesy Ursuline College Historic Costume Study Collection*

Top right: Back view.

Bottom left: Detail.

Center: Cape in silk faille. Hand embroidered ornamentation in jet beads on silk braided appliqué, c.1880s. *Courtesy Anna Greenfield*

Bottom right: Detail.

Top: Cape in silk faille and netting. Hand crocheted and embroidered with jet beads, c.1880s. *Courtesy Emma Lincoln*

Bottom: Detail.

Top: Cape in silk faille. Hand tambour beaded in jet with a 5-inch fringe, c.late 1800s. *Courtesy Emma Lincoln*

Bottom: Detail.

Top: Cape in silk faille. Hand beaded in jet with 4-inch fringe, c.late 1800s. *Courtesy Emma Lincoln*

Bottom: Detail.

Where Are Bead Making and Bead Work Being Done Today?

There is a world-wide fascination with beads and beadwork today. From the 1960s, when the "flower children" of the boomer generation slipped on those beaded necklaces, the craft of beads and beading has grown in unprecedented popularity. Ranging from traditional to contemporary materials and employing both ancient and modern techniques, bead making and beadwork can be found everywhere.

During the late 19th century and the first half of the 20th century, bead embroidery was worked principally by French and Eastern European embroidery artisans on couture fashions made for exclusive clientele. Obtaining the intricate beadwork and the exquisite embellishment of *haute couture* was an impossibility during the Second World War, and it was too costly for Post War ready-to-wear fashions. By the '60s, however, beading was beginning to be featured as trim for evening wear. Apparel accented with beading could be purchased in ready-to-wear shops. In the late '70s and through the '80s, cheap Far Eastern labor and the lure of the television soaps, *Dynasty* and *Dallas*, brought "flash and dazzle" beading. The over-beaded dress, suit, and gown became daytime as well as evening fashion.

Bead embroidery for the fashion industry continues to be done in the Far East. Beaded laces and fabrics are produced at competitive prices in Thailand, Malaysia, Phillippines, Japan, China, and especially in India. Expert bead workers in Bombay, on the western coast, and Bangalore, in south central India, work the artistry of bead embroidery for the European and American couture fashion designers.

India has a long history in bead making and working, beginning c.1000 B.C. Much later, from the 16th century to the 18th century A.D., under Muslim rule, bead and jewelry making flourished along with gold bullion embroidery. Mark Brower, an Associate Designer with Vera Wang, Ltd., New York, explains that in the Muslim tradition most of the bead embroidery today is done by men, especially in central and south India. In the north, more women are found working as embroidery workers. According to Laura Moffhet of Mary McFadden Couture, New York, couture bead workers in India hold a prestigious position and work in relatively good conditions. Mary McFadden gowns have been embroidered in India from her first collections in the mid 1970s.

Many of the skilled bead embroidery workers in the United States are originally from Thailand, Cambodia, and other Far Eastern countries. Elizabeth Courtney Costumes of Studio City, California, creators of the spectacular Bob Mackie beading, have a team of expert craftsmen to work the exquisite beading of his couture pieces.

European fabrics, laces, and bead embroidery today remain the most costly and, therefore, the most exclusive.

Bead necklace belonging to rock star Janis Joplin, c.1970. Presented to the Rock and Roll Hall of Fame & Museum, Cleveland, Ohio, by *Rolling Stone Magazine. Courtesy Rock and Roll Hall of Fame & Museum, Cleveland, Ohio. Photo Tony Festa Photography.*

Beads on the back too! Bob Mackie gown covered with hundreds of thousands of crystal bugle beads, embroidered by hand at Elizabeth Courtney Costumes, Los Angeles, California, c.1985. *Courtesy Carole Carr*

Chapter 2
Types & Techniques

Are There Different Kinds of Beads?

Prehistoric man used natural objects for the first beads: berries, seeds, nuts, shells, wood, bones, and stones. As the culture evolved, beads were made from clay, metal, eventually glass, and during the 20th century, plastic. Into the 21th century, the new millennium, all of these materials are used to create ornamentation throughout the world. Bead materials include: seeds, beans, nuts, shells, egg shells, wood, bones, horn, coral, pearls, mother-of-pearl, and natural and semi-precious stones. The list goes on: cloisonné, cotton and ceramic beads, faience, foil, and knot beads, metal beads, porcelain and plastic beads—and, of course, glass beads. Using the definition "a roundish, perforated object," the only limit might be individual creativity, not material.

Wound, Drawn, Molded, and Blown

Glass canes were originally made by a gaffer who would dip a long metal rod (mandrel) into a crucible of molten glass, pull out a glob (gather) of red-hot, viscous glass, "settle" (marver) it against

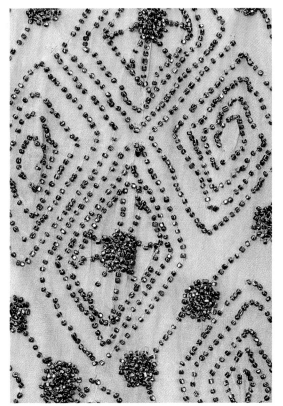

Steel seed beads embroidered on silk, c.1920s. *Courtesy Ursuline College Historic Costume Study Collection*

short mandrel (metal rod, originally treated with carbon) is held in one hand and a glass cane in the other. The Venetian bead maker of 500 years ago would use the flame of an oil lamp connected to bellows to heat the mandrel and melt the end of the glass cane. Today, bead-makers use an oxy-propane glass-working torch to create an even heat, and they treat stainless steel wire rods with a ceramic release agent.

As the end of the cane melts, it makes a small gather of semi-molten glass that is touched to the mandrel. The mandrel is then turned to wind the melted glass around the wire, forming the basic bead. An instrument such as a carbon paddle is used to flatten and shape (marver) the bead. The wire rod must be rotated constantly to keep the glass bead at the correct temperature. This process is repeated by adding tiny drips from glass canes of other colors in the most delicate technique to provide color and form and to create faces, fish, flowers, vegetables, and other tiny works of art. One bead is made on one rod at a time. The bead, when complete, is cooled. Then the bead is pulled off the wire rod and placed in a kiln at 975 degrees Fahrenheit for an hour, a process called annealing. Patience and practice are required: it may take months to perfect a technique and hours to produce one bead.

The method for making drawn beads is ancient. Here, a glass blower takes a gather of molten glass on his blow pipe from the crucible in the furnace. After marvering it, he blows a bubble.

His assistant takes a gather on the end of a metal rod and they both reheat their glass in the furnace. Then the assistant attaches his gather to the bubble. Both of the men move quickly away from each other, pulling the glass bubble

marble, reheat it, and swing it to an assistant who would catch the lump on his metal rod.

The assistant would quickly walk away, pulling the mass into a glass cane that would solidify before it touched the floor. The gaffer would judge how quickly and how far the assistant would move to produce the precise diameter of the cane, which he calculated by sight. After cooling, these glass canes were then cut into more manageable lengths. Today, this process is done by machine.

The current standard method of creating wound or lamp beads was developed in Venice in the early 16th century. A

into the form of a long, thin tube, sometimes more than one hundred feet long. The "bubble" is stretched to become the "hole" in the tube. The tube is cooled and then broken into meter lengths later to be cut into small beads. This is the monochromatic drawn glass bead, but variations and decorations can be made by layering different colors on the basic bubble. The beads are sometimes tumbled in a powder over heat to round the rough cut edges. Seed beads and bugles are made this way.

Blown beads are made by making a small tube as above and then inflating it into a bead. Each bead is made individually, or the glass blower can use a mold with a series of compartments to make several beads, which can then later be cut apart.

Molded beads are made by shaping the hot glass between two halves of a mold. The mold leaves a seam line on the bead that needs to be ground off by hand or tumbled to smooth the lines.

Although machines have been built to do much of the work for the tiny seed beads, the more complex beads are still made mostly by hand. In the case of vintage beaded apparel, each bead represents hours of hand labor spent to produce beads for the embroiderer to work into marvelous patterns.

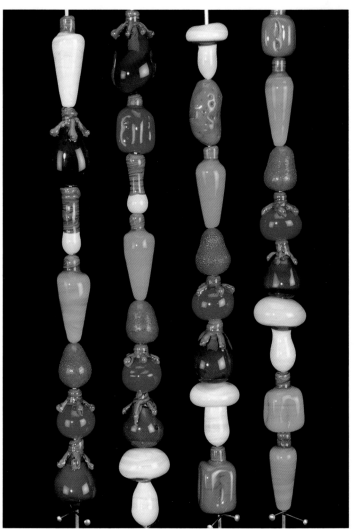

Right: Lampwork wound beads in the shape of vegetables, c.1990s. *By Sandy Osborn*

Below: Wound, blown, molded, wooden, stone, and plastic beads: antique and contemporary. *Courtesy Shirley Friedland*

Which Types of Beads Are Used to Ornament Apparel?

Glass Seeds and Bugles: drawn beads

In general, the types of beads used to ornament apparel are the small, glass, drawn beads, called seeds and bugles. These tiny glass orbs and tubes, which delight with their reflected and refracted light, were sold by the pound as the glass bead trade began. They came to be known as pound beads. They were produced in the glass factories of the Venetian Republic by the Margaritieri division of the glass making guild (Christalleri). Both seed and bugle beads come in many sizes and finishes.

The seed bead or rocaille, terms used in the trade, refer to any small glass bead that is used for beadwork. These beads may also be described as "sand" and/or "micro" beads and are processed by tumbling, which rounds and smoothes them after they have been cut from the glass tube.

The bugle bead, as it is defined today, is usually a small bead shaped like a tube. The etymology of this word is not clear. In history, the bugle bead referred to both round and long shaped glass beads. Glass beads began to be used for embroidery in the late 16th century, and there are references to "lace bugles," probably used for work on fabric. The "bugle" came to be used for what the Italians call *tubetti* and the French, *longue perle de verre*. It is left "unpolished" as it appears when sliced from the glass tube.

Definitions of terms used to describe these beads are sometimes confusing, because the vocabulary of bead embroidery may be different from one beader, supplier, or historian to another.

Most of the beads photographed and described in the following catalog are seed and bugle beads, which may have a variety of finishes or treatments.

Cut or Faceted Beads

Faceting, a word derived from the French word for "face," is the cutting of flat planes on a rounded bead to produce a greater play and reflection of light. Until the 18th century, few beads with facets were found.

Names given to faceted beads are: one cuts, two cuts, seed beads with one or two flat facets; three-cuts, irregular

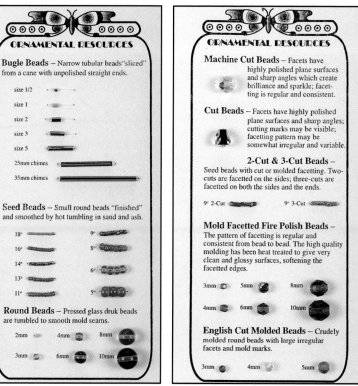

Sampler of bugle and seed beads showing sizes. *Courtesy Ornamental Resources, Idaho Springs, Colorado.*

Sampler of cut beads. *Courtesy Ornamental Resources, Idaho Springs, Colorado*

cuts all around the bead (*Charlottes* is a term that is sometimes used interchangeably for beads with two or three facets); and hex cuts, originating from a cane with six sides. *Tosca* is the name given to beads that are smooth on the outside, but square-cut in the center to catch the light.

Pony or Crow Beads

These are larger round beads that have been used in bead embroidery around the world and can be found on some of the apparel that appears in this book. Pony beads were named so in the 19th century, because they were brought by pony to be traded with the American Indians. They are also known as Crow beads, after the Crow Indian bead workers who favored these larger beads for the edges of some of their embroidered articles.

Delica and Antique

These are the newest beads on the block and the most popular for contemporary beadwork and embroidery. Delica and Antique are brand names for beads developed in Japan that are uniformly cylindrical with large holes and can be worked to produce an even, exact pattern due to the consistency of their design.

Molded or Pressed

Molten glass cut from the cane is pressed into shapes such as hearts, stars, and flowers and used in bead embroidery.

Color Effects

Opaque: no light can pass through; "solid" in color and therefore the thread cannot be seen through the bead.

Transparent: light passes through; the thread is seen, even with a colored bead, and the color of the thread will affect the design.

Translucent: between opaque and transparent; the color of the thread may affect the design.

Iridescent or *Aurora Borealis* (AB): an effect on the surface of the bead that creates a "rainbow" of colors.

Metallic: a shiny surface that looks like metal, usually produced by color that is baked on.

Sizes

The method used to measure seed beads is based upon the sizing of the glass canes used to make the beads. The canes are measured in degrees or increments of aught, meaning zero: the smaller the number, the larger the bead. Pony beads are usually sizes 6-aught to 9-aught, or 2mm to 5 or 6 mm. Seed beads are smaller than 2mm, and usually sized from 10-aught to 18-aught. Vintage beads can be found that are even smaller, as minute a size as "20-aught" and beyond. It is difficult to imagine the needle that would be used for such beads. Each country producing seed beads calibrates differently. So a size "11-aught" from one country origin may be different from a size "11-aught" from another country.

Bugle beads are measured by length. They are usually equivalent to the size "11-aught" seed bead. The bugles best used for embroidery are sized from "1/2" to "5".

The largest of the beads described above can be made in plastic, and, although not as fine and beautiful as glass beads, their light weight can be an asset.

Are There Different Methods of Beading?

Today, as in the past, seed beads can be strung, knitted, crocheted, woven on a loom, or woven off-loom using just a needle to create a textile made of beads. Along with bugles, sequins, artificial jewels, and passementerie, seed beads can be attached to material in beautiful designs for bead embroidery.

Sequins or **paillettes** are important to mention here, because they are combined with beads in much of the beaded apparel of the 20th century and in the apparel included in the following catalog. **Paillettes** originally referred to circular discs, measuring 2mm to 8mm, stamped from metal with a hole in the middle. In Elizabethan times, there also was reference to a "spangle," which, like the paillette, was also circular and metal (sometimes gold) with a center hole; however, these were flat. "Sequin," originally a term for certain Italian coins of the 13th century, has become the general name for anything that resembles these descriptions, regardless of the size, shape, or number or position of its holes. Today, paillettes normally have one hole at an edge; this allows the disc to flap freely.

At the end of the 19th century, new materials like colored animal gelatin and lacquered rubber replaced the metal that formed the earlier paillettes. Both types of paillettes, however, proved unsatisfactory. In *Haute Couture Embroidery, the Art of Lesage*, Palmer White relates a humorous story of a grande dame who attended the opera in a gown covered in gelatin paillettes, only to become overheated and find her dress melting into consommé!

Faux jewels also have a history of use in bead embroidery. Colored glass cut to resemble natural gems and semiprecious stones was a popular substitute for the middle classes through the 19th century.

The now familiar term, rhinestone, came into English from the French at the end of the 19th century.

Passementerie is the name given to trimming made of gold or silver braid that came to Europe from the Orient. Italy became the main producer, but, in the middle of the 16th century, a guild was established in Paris. The technique was perfected by the French in the 18th century. Later, the Victorians encrusted the braiding with beads to decorate apparel.

Cuffs or bracelets; hand embroidered on velvet in seed beads in a floral design with a white butterfly, c.1850. *Courtesy Veronica Trainer*

Opposite page:
Top left: Tambour tools: holders and hooks, c. early 1900s. *Courtesy Russell Trusso Atelier*

Top right: Tambour beading on silk with steel beads in diamond pattern; clusters are beaded with a needle, c.1920s. *Courtesy Ursuline College Historic Costume Study Collection*

Bottom left: Tambour beading in jet and black glass beads, c. late 1800s. *Courtesy Ursuline College Historic Costume Study Collection*

Bottom right: Reverse side of hand tambour beading on net, c. early 1900s. *Courtesy Ursuline College Historic Costume Study Collection*

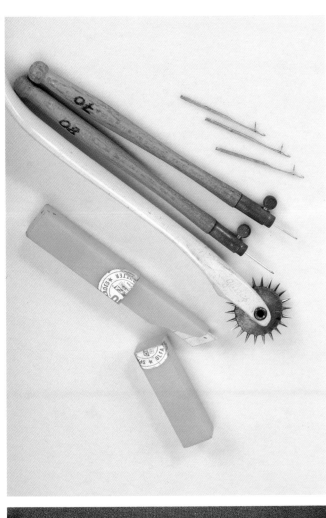

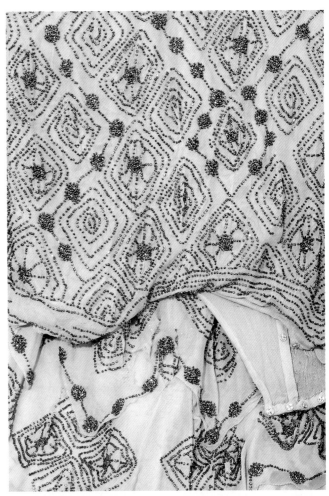

Embroidery Techniques

One bead at a time is how it was done and how it continues to be done in most of the industry. When explorers noted that American Indian women adopted an innovative technique of stringing five or six beads together before sending the needle back into the fabric, the stitch was given a derogatory name of "lazy squaw" stitch. Today, there are bead embroidery books that provide easy-to-follow instructions and illustrations for attaching a singular bead and/or groups of beads in a variety of stitches.

The basic technique for embroidering has changed little. Publications on needlework and bead embroidery, up to five hundred years old, record methods and stitches that are similar to current instructions. The roots of the chain stitch can be traced back thousands of years to China where it was used in silk embroidery. The stitch found its way to India, where the embroidery known as "shisha" mirror-work and a specialized hook used to make the stitch were developed.

Tambour beading, a chain stitch embroidery using a tambour hook, was noted in a manual of Charles de Saint-Aubin of France, who was referred to in the later part of the 18th century as "Costume Designer to the King" (Dessinateur du Roi pour le Costume). Around 1750, the chain stitch had already begun to supersede the needle-sewn-one-at-a-time stitch. The hook in French was called *aiguille* a *chainette* or *un crochet*. In English, it is called tambour hook, named for the French round frame used to hold the fabric taut.

The technique of attaching beads with the tambour hook did not appear until the end of the 19th century. About the same time that the demand for beaded fabric brought about the development of the Cornely machine for sewing beads, Louis Ferry in Luneville, France, developed the use of the tambour hook for beading, *la broiderie perlee de Luneville*. It made beadwork faster, neater, and less expensive to produce. Tambour beadwork and the machines adapted to attach beads made opulent beadwork of the 1920s possible.

The following beaded yardage is *Courtesy of Virginia Marti Fabrics, Cleveland, Ohio.*

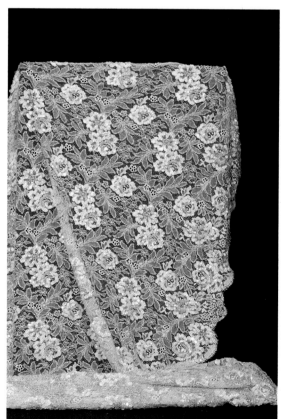

Cream lace yardage embroidered with matching luster finish bugle beads, and white and clear sequins.

Detail.

Left: Pink lace yardage embroidered with silver-lined bugle beads, rocailles, and tiny clear sequins.

Right: Detail.

Left: Beige lace yardage embroidered with luster finish seed and bugle beads, round and rice shaped pearls, and iridescent sequins.

Right: Detail.

Left: Bronze lace yardage embroidered with matching micro seed beads and bronze sequins.

Right: Detail.

Left: Lilac and silver lace yardage embroidered with matching seed beads and silver sequins.

Right: Detail.

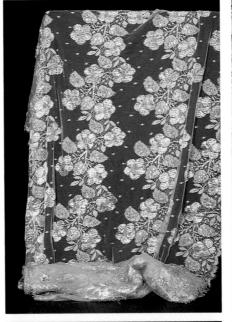

Left: Lavender lace yardage embroidered with matching bugle beads.

Right: Detail.

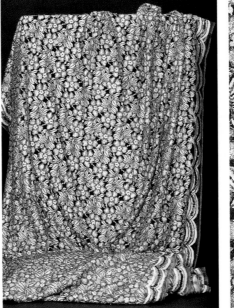

Left: Gold and copper lace yardage embroidered with gold balls and seed beads.

Right: Detail.

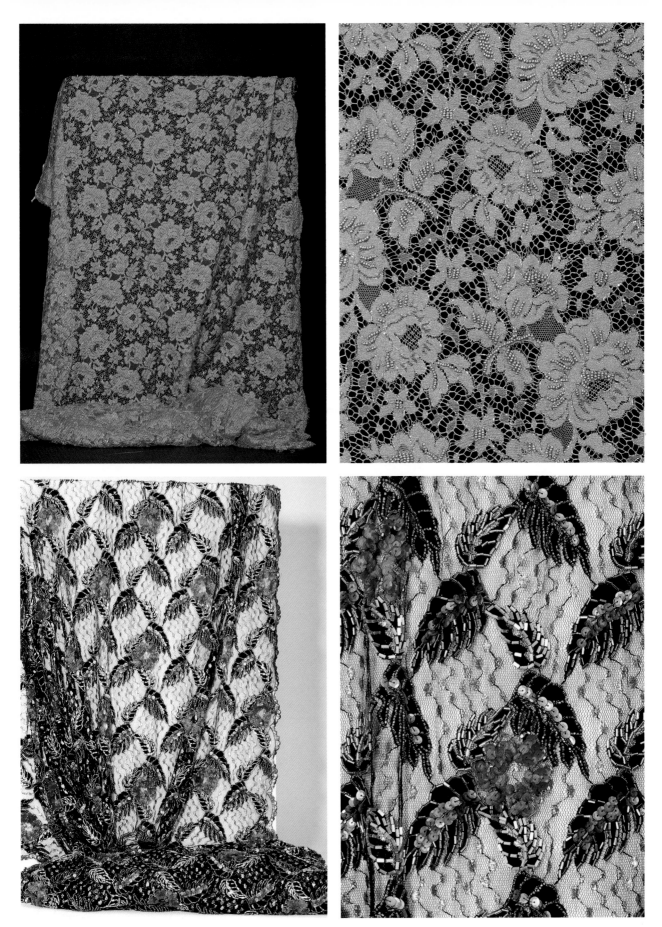

Top left: Fuchsia lace yardage embroidered with matching seed beads.

Top right: Detail.

Bottom Left: Black lace yardage embroidered with iridescent green seed beads, green, gold-lined, gold luster finish bugle beads, and fuchsia sequins.

Bottom right: Detail.

How Is Hand Beading Identified?

Handwork can usually be identified by uneven stitching and knotted threads on the reverse side. Hand tambour beading is not always easy to identify due to the even pattern of the chain stitch. However, with careful examination, the different tension and spacing between the chain loops can be discerned.

Can the Place of Origin Be Determined by Identifying the Beading Method?

The beading method alone does not determine the place of origin. At times, along with examining the "style" of the beading patterns and/or the use of certain embellishments, the country where the embroidery was produced can be identified.

Can Identification of the Beading Method Aid in Dating a Piece?

In the case of "one-at-a-time" hand beading, dating a piece can be difficult without more information. Dating tambour beading is easier—nearly always after the 1880s.

What Is the Lesage Method?

Haute Couture Embroidery: Tissus Albert Lesage & Cie

In the 19th century, nouveau riche bourgeois women flaunted their husband's wealth by wearing furs, jewels, and sumptuous gowns. Even though machinery of the Industrial Revolution provided the new wealth, machine-made apparel was rejected. They desired, demanded, and could afford elegance unavailable from fashion sources. *Haute couture* began with the Englishman Charles Frederick Worth, who emigrated to France in 1845. He was the first fashion designer to dictate taste and design; he gave dressmaking "snob appeal" and was the first to put a label on his clothes.

Albert Lesage was born in Paris in 1888. His talents for languages and especially for sketching served him well in his early years. In 1910, at age twenty-two, he was a broker-consultant, a *commissionnaire*, for the French couture houses. He also kept his American clients of New York, Boston, and Chicago informed of new developments in fashion, and he advised them on what to buy.

After being a prisoner of war for four years during World War I, Lesage returned to Paris and to the *commissionnaire* in 1919. His sketches found their way to Mr. Field, president of Marshall Field and Company of Chicago, who hired him as the manager-designer of the women's made-to-order dressmaking department. In Chicago, his talents were not well served by this method of attaching a Marshall Field label to a product of an unknown designer: Chicago's elite could purchase a labeled Paquin ensemble down the street. So, in 1922, Lesage returned to Paris.

The French fashion industry was changing—women in western Europe and in the United States were changing. The small embroidery firm of Michonet was well known in Paris for supplying embroidery and passementerie to European royalty and to the grand designers of *haute couture*. Lesage purchased Michonet's business including the archives of drawings, designs, and meticulous accounts dating from the earliest years with Charles F. Worth.

In 1924, as Lesage was ready to take ownership of the company, the French franc was devalued, and the rush was on by the "society" of North and South America and all of the major cities in Europe to purchase Paris fashions. Lesage et Cie prospered during the next five years, producing elaborate embroidery for the great fashion houses and creating over 1500 pieces of exquisite embroidery for Madame Vionnet alone. The Great Depression and the Second World War interrupted the fashion industry, but, throughout those times, Lesage et Cie survived. Today, *haute couture* fashions embellished with embroidery by the house of Lesage are works of art displayed in museums around the world.

Logo of Albert Lesage & Cie on portfolio. *Courtesy Doreen Leaf-Hund*

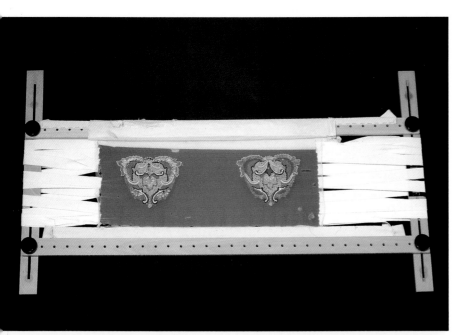

Frame and tambour bead embroidery in progress using balsa wood beads, amber bugles, black faux jet, raffia, silk thread, and wood patterned sequins. Project from the Lesage School, Paris, France. *Courtesy Doreen Leaf-Hund*

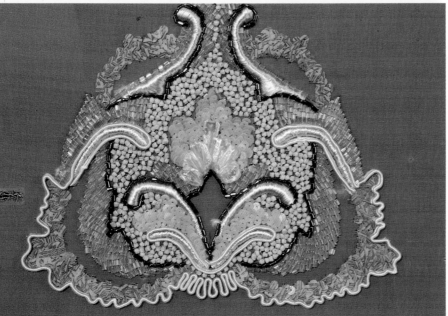

Front view.

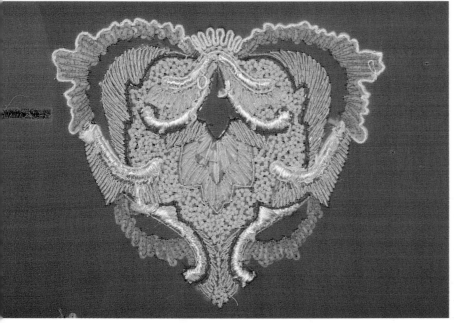

Back view.

33

Chapter 3
Condition & Care

How Important Is Condition?

Most expert collectors of vintage fashions agree that condition is extremely important. By necessity, collectors of bead embroidered apparel and accessories may be more forgiving when evaluating condition. One wearing of an exquisite, yet fragile, beaded gown can cause sufficient damage to relegate it forever to the back of the closet! Invariably, items found at reasonable prices in vintage clothing specialty shops will have damage.

Most experts agree that some damage is acceptable. Deciding whether the condition of an item is worth the price and the time required to repair it has a great deal to do with the intended use of the piece and the part it will play in the collection.

How Is Condition Evaluated?

Bead embroidery is many times more likely to have problems with condition as other vintage apparel of equal age. First, beads, whether glass, natural materials, or plastic, are weighty, especially in quantity. They pull at the threads and weaken the fabric. Second, the beads themselves have edges that wear on the threads. Third, especially since the 1920s, beaded fashions have been created in fragile, diaphanous fabrics of net, delicate silks, and cottons, and, today, delicate man-made fabrics. These fabrics cannot take the weight of the beads for long. Fourth, the embroidery thread pulls, ages, weakens, and breaks, popping the tiny seed beads as the tambour chain stitch unravels.

Some references suggest avoiding bead embroidered apparel all together because of the predictable loss of beads and overall damage. For bead lovers, this would be impossible, despite that they will shed tears over the inevitable loss of beads.

Examine the fabric first. Hold it up to strong light. Look for tears, holes, and areas that appear weakened by thinning. Then look carefully at the beading and the pattern. Are there missing rows of beads or hanging beads?

A dress that has been stored for years may appear "strong" and in excellent condition at first sight. When handled, even gently, for example in photographing the item, the fabric may begin to disintegrate. The authors experienced this with more than one of the gowns shown in the following catalog.

What Types of Damage Are the Most Difficult to Repair?

Splits or shredding to fabric are the most difficult problems to repair. This damage from aging and lack of care is called "shattering." The material disintegrates along the warp threads, and the condition is irreversible with no shortcut to stop it. An expert restorer would use a "brick stitch" basting pattern and spend hundreds of hours hand sewing the entire surface of the damaged fabric onto another suitable fabric for reinforcement. Such a restored garment should be carefully stored and never worn.

A beautiful beaded design that has large spaces of missing beads is also very difficult to repair. Learning the bead embroidery method would not be difficult, but finding the matching beads would take time, research, and may even be impossible.

Soiling from perspiration, especially on silk, is almost impossible to clean. To wear the piece, restorers suggest attempting to match the fabric and cutting away and replacing the damaged areas.

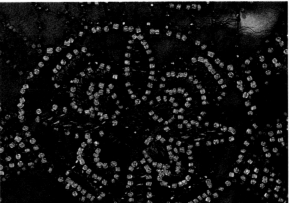

Top left: Bodice in black silk showing damage due to shattering. Bead embroidery in dyed pink, black seed beads, and black silk thread, c.1920s. *Courtesy Ursuline College Historic Costume Study Collection*

Top right: Detail.

What Types of Damage Are the Least Difficult to Repair?

Split seams are relatively easy to repair provided that there is enough material to allow for a smooth, strong seam.

Missing beads from fringes are much easier to replace than damaged design work. Fringe work, stringing rows of beads, is generally easier to accomplish than applying beads in a pattern. Often it is not necessary to match the exact bead to complete a damaged fringe and achieve a satisfactory appearance.

Left: : Bodice in black lace with jet bead embroidered appliqué, c.1800s. Missing strands of fringe in faceted jet beads can be replaced. *Courtesy Ursuline College Historic Costume Study Collection*

Right: Detail.

Can and Should Delicate Antique Fabrics Be Reinforced or Repaired?

Mending or reinforcement of delicate vintage pieces should be done with care by someone familiar with hand sewing techniques and restoration work. There are a number of professional restorers or conservators who would be able to assess the damage, and suggest and complete a restoration project. The cost involved will depend on the hours of handwork that is required.

In addition, as collecting vintage clothing gains importance, a number of books have become available with excellent instructions, illustrations, and resources to guide a collector through the process of repair and mending.

But beware, a poorly repaired garment will make matters worse, causing the beauty of an item to diminish along with the item's value, and calling attention away from the piece and onto the damage done to the piece.

Next page:
Top left: Peach silk dress showing damage to straps due to shattering. Because the dress hung for years on a wire hanger, the weight of the dress contributed to the deterioration of the shoulder straps. Tambour bead embroidery is in crystal bugle beads and peach pearls in a "feather" pattern. c.1920s. *Courtesy Ursuline College Historic Costume Study Collection*

Top right: Hand stitched repair to shoulders using commercial ribbon in matching color.

Bottom left: Cream lace over yellow silk. In otherwise very good condition, the dress has been altered with the gathered cummerbund in a synthetic fabric designed for lining; the alteration detracts from the original beauty. Hand tambour embroidered in gold-lined rocailles, bugle beads, and gold thread. c.1915. *Courtesy Ursuline College Historic Costume Study Collection*

Bottom right: Detail.

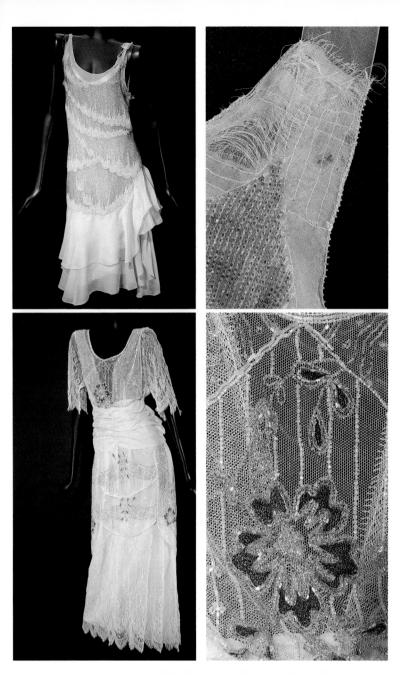

the embroidery. Wear full undergarments to protect the fabric from perspiration and body oils. Do not apply perfume that can come in contact with the clothing.

As with all clothing, be attentive to small problems and damage so that these can be repaired before they become large, irreversible problems. In the early part of the 20th century, a lady would have a seamstress and beader to repair the dress immediately. If you decide to wear your vintage collection, you will most likely need to be its caretaker.

What Preventive Measures Can Be Taken To Protect Delicate Beadwork?

Other than wear, exposure to light and dust are the main causes of damage to vintage clothing. This applies to other beaded articles, as well.

Cleaning

Whether antique or modern, delicate and fragile beadwork should be cleaned professionally by a reputable firm experienced in beaded fashions or bridal gowns. Visit a shop that specializes in beaded gowns and ask the sales professional for the names of local cleaners that they might use.

Vacuum cleaning to remove dust is possible if the bead embroidery is on a strong fabric. Use the small nozzle of the vacuum covered with a piece of monofilament filtering material to prevent beads from being vacuumed up.

A sable paint brush can be used carefully on the beading to lift the dust before vacuuming with a filtered attachment. Avoid vigorous brushing or wiping beaded items; this can break the weakened threads and result in damage.

Bead embroidered items should not be immersed in water or washed. Some beads are lined with a color and, if washed, they will return to clear crystal, destroying the original design.

Instead, press with a cool iron, or very carefully steam the piece, without touching it, using a protective cloth between the iron and the item.

Storage

All bead embroidered apparel should be stored flat, never hung. Wrap the piece with acid free tissue paper or prewashed unbleached muslin, taking care that the beaded areas are not touching other beading; this will avoid tangles. Pillow cases can also be used for storing. To avoid folds in the fabric, crush the acid free tissue paper and stuff the bodice, sleeves, and the skirt till they are full. Make sure to smooth out the creases and equalize the weight of the item. Find a storage space for vintage fashions other than the attic or the basement, a place without extremes in temperature or humidity.

Before storing vintage apparel, remove plastic bags from the cleaners. Plastic bags will cause the fabric to retain moisture and discolor or mildew. Remove all pins, staples, or other metal attachments that may cause rusting.

Before tucking away your vintage fashion, take a second look. This dress or accessory, whether beaded by hand or by machine, is not replaceable.

Should Vintage Fashion Be Worn?

Historical costumes should never be worn for any occasion or purpose. This refers to the costumes carefully guarded in low light controlled environments in costume museums. It also refers to some of the *haute couture* pieces being sold at auction.

The definition of vintage fashion can be as "new" as the 1970s, as the term is currently used. Popular fashion currently includes the wearing of vintage clothing. Therefore, go ahead and wear, but wear with care.

Heavily beaded dresses, regardless of their age, will suffer the most, even from one wearing, especially one that includes dancing. Resist wearing the magnificent bead embroidery of the 1920s. Admire it. Treasure it. If the clothing is dated 1900 or older, again, admire and treasure it, but don't wear it.

In all cases, do not sit in beaded dresses. Avoid wearing jewelry that can tangle with the beads and threads and pull on

Chapter 4
Collecting & Value

How Important Is Quality? How Is Quality Determined?

"Quality" vintage apparel will have and always retain a higher value. A quality piece of bead embroidered apparel will not only have beautiful and unusual patterns of beadwork, it will also have fine workmanship, excellent fabric, and good lines, color, and design, with possibly a recognizable label.

The "eye" rather than a label will identify the quality. Many of the vintage bead embroidered articles found in vintage shops have no designer or store label. Look carefully at the material. Is it being used today? The beaded trim may be vintage, but the dress or ensemble itself may be a costume reproduction from a theater. Train your eye by studying examples in books on the history of fashion, or by visiting museums, vintage clothing shops, fashion shows, and auctions.

How Is the Value of The Beaded Article Determined?

Condition, quality, design, rarity, desirability, and, once again, condition determine the value of bead embroidered clothing and accessories. Desirability refers to trends or fads in the market. The 1997 highly successful charity auction of Princess Diana's couture gowns spotlighted numerous vintage fashions and will have a lasting effect on the market.

Desirability also refers to the rule of thumb 'collect what you love or desire.' At one time, few collectors would bother with the difficulties of beaded apparel. In the past fifteen years, beadwork and beads themselves have grown in popularity and in desirability.

Just so many dresses were made in 1920s. The flapper era was relatively short, not even lasting the full decade, so finding a beaded flapper gown in good condition outside of museums is difficult.

These factors affect the value of the beaded article, but there is no set method of pricing. Irresistibility also affects the price a collector is willing to pay for an article.

Is the Size of the Bead a Factor in Determining Desirability?

Most of the bead embroidery on fashion in the 20th century is worked in beads that are of similar size. Finding an item embroidered with tiny, micro beads will delight any collector. The smaller the beads, the more difficult and time consuming the process was to create the piece.

Are Pieces of Bead Embroidery Worth Collecting?

For a true bead lover, the answer is an easy "yes". Bits and pieces of bead embroidery can be appreciated for design, color, and workmanship. Pieces of beadwork are beginning to show up in antique shops, because too often only pieces have survived. Makers of wearable art sometimes incorporate antique remnants into contemporary creations.

Cream silk net embroidered with matching silk thread; tambour beaded with pink and cream sequins, silver-lined bugles and rocailles in a floral design. Sample is from the over skirt of a dress. c.1920s *Courtesy Ursuline College Historic Costume Study Collection*

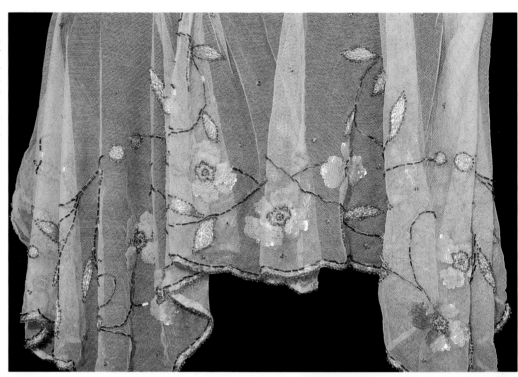

Detail.

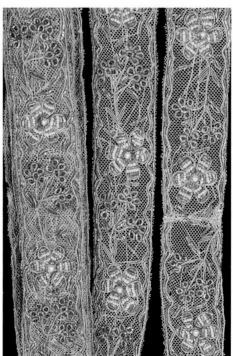

Cream lace passementerie machine embroidered with gold thread, green and red seed beads with pearls, c.1960s.

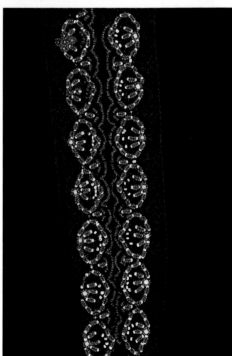

Black lace passementerie hand embroidered with red and white seed beads, c.1980s.

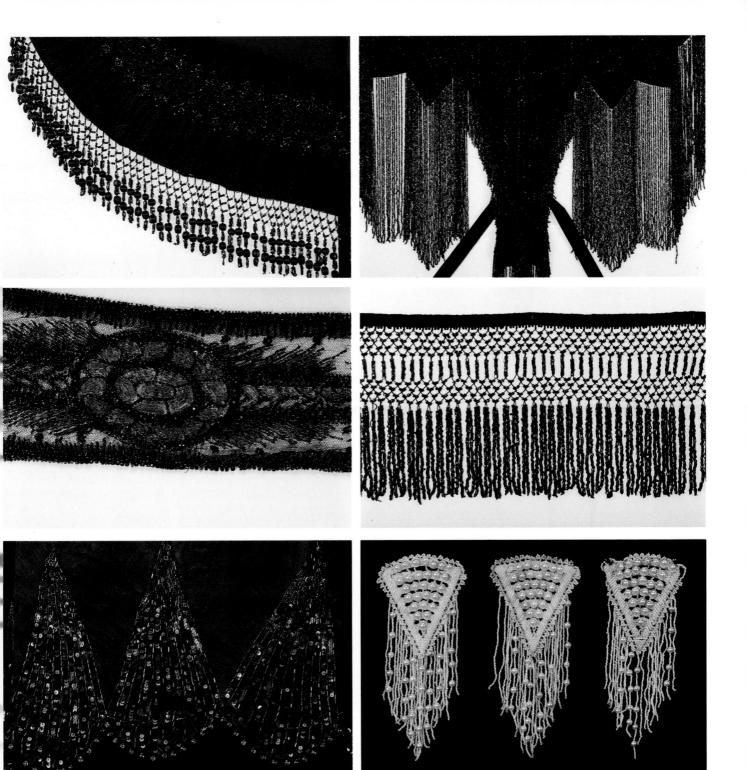

Top left: Trim of mourning cape; black jet hand embroidered on braiding and on thread lattice work. Fringe is in jet and accent beads made from silk thread around fabric balls. c.1880s. *Courtesy Ursuline College Historic Costume Study Collection*

Center left: Black net passementerie hand embroidered with jet, black glass beads, and tiny gold sequins, c.1920s.

Bottom left: Black silk, tambour beaded with iridescent seed and bugle beads and tiny sequins, c.1930s. Sample is from the skirt of a dress. *Courtesy Ursuline College Historic Costume Study Collection*

Top right: Black 12-inch fringe and hand bead work in jet and black glass beads on a mourning cape, c.1880s. *Courtesy Ursuline College Historic Costume Study Collection*

Center right: Black 4-inch fringe in jet beads, c. late 1800s. *Courtesy Ursuline College Historic Costume Study Collection*

Bottom right: White pearls and luster finish seed beads hand embroidered on tape appliqué, c.1930s. *Courtesy Sandy Osborn*

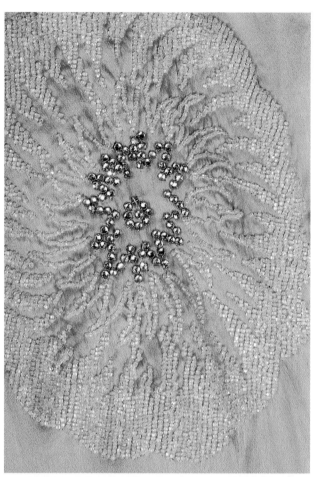

Top left: Black net appliqué hand tambour stitched with jet and seed and bugle beads, c.1890s.

Top right: Gray silk hand tambour beaded with crystal seed beads and set rhinestones in a floral pattern, c.1930s. Sample is one of nine floral embroideries on the skirt of a dress.

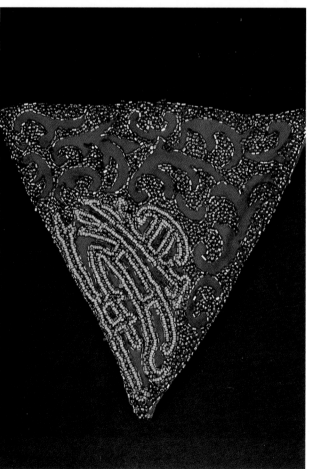

Bottom left: Sleeve with hand tambour stitched pink and white seed and bugle beads, c.1920s

Bottom right: Copper colored crepe appliqué hand tambour stitched with matching bugle beads, turquoise and charcoal hematite seed beads, c.1940s.

 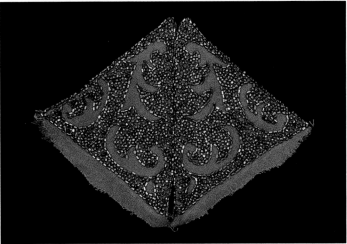

Cuff with hand tambour embroidery on silk; in satin finish and silver-lined bugles, c.1885. *Courtesy Ursuline College Historic Costume Study Collection*

Copper colored crepe appliqué hand tambour stitched with matching bugle beads, c.1940s.

Where Can Beaded Fashion Be Found?
Are There Still Bargains Waiting To Be Discovered?

Following the charity auction of Princess Diana's gowns in 1997, auction houses began to establish or enlarge departments of antique clothing and apparel. William Doyle Galleries of New York was one of the first to recognize the importance of vintage fashion, and they conducted auctions of couture and antique apparel much earlier.

Beaded fashion can be found in vintage apparel shops, thrift stores, and estate sales. Occasionally, garage sales, where bargains may still be found, might hold a couture design that was regarded as "old clothes."

What Should Collectors Look For in Forming a Collection?

As with all collections, collect what you love! But be informed. Begin by visiting historic costume museums. Research the sales catalogs of the auction houses and look for books on vintage clothing and beadwork. The collection might be based on the collector's personal taste, period, designer, or a combination of criteria, but what the average collector should look for first is condition.

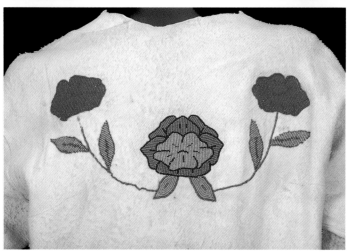

Tan two piece contemporary costume with antique black lace passementerie accented with black seed and bugle beads. *Courtesy Ursuline College Historic Costume Study Collection*

Top left: Natural hide Alaskan Indian dress with floral appliqués on cardboard backing in opaque seed beads. Example of modern work in the manner of early Indian dress. *Courtesy Frances Dickenson*

Center left: Back view.

Bottom left: Detail.

What Should Collectors Avoid?

What the collector selects or leaves behind directly relates to personal preference and price range.

It is advisable to refuse articles of severely disintegrating fabric. Items from the Victorian Era are sometimes found to be "flaking," leaving a residue of black powder. This condition is progressive and irreversible.

Alterations or repairs to a garment that affect the integrity of the original design, the piece's historical significance, or ones that cannot be successfully reversed should be avoided.

Large stains or faded areas that cannot be repaired are also reasons to resist purchase.

Again, it is the condition of the article that should be closely examined before deciding on the purchase.

What Are Some Ways in which to Display Beadwork?

In addition to wearing vintage fashion, bead embroidered items and apparel can be displayed as decorative art: frame pieces of bead embroidery, even an entire garment, and hang on the wall for display; hang a 1920s dress on a padded hanger from the corner of a screen or mirror, or from a hook for display; or position a mannequin modeling a vintage beaded garment in a bedroom, dressing room, or even a living room for display.

Look for ways to allow the collection to be seen and admired by the viewer everyday, but be careful that dust is not allowed to accumulate. Dust particles are abrasive and can weaken the threads that hold the beads.

What About Purses?

Beaded purses, whether embroidered, knitted, woven, or crocheted, have a long, rich history and merit a study of their own. There are many publications that provide specific history and information for collectors, and the authors of this book will be publishing a companion volume entitled *Beads on Bags*.

Is the Antique or Secondary Market Comparable to New Retail Prices?

Generally, new retail prices for comparable items will be the highest. Secondary market prices for vintage items can be higher or lower than original retail prices, depending on the age, quality, and designer attribution. Relatively recent items with very high original retail prices will show the most depreciation in secondary markets. Top designer labels will cause an older item to retain or increase its value.

The price range suggested at the end of captions is intended as a guide. It is based on retail prices in secondary market shops, auctions, etc., with a few exceptions—some captions include original retail only, and price ranges for recent beaded shoes are closer to original retail, since shoes are not yet frequently found in vintage shops. Prices are for items in excellent condition (even if the example shown is not). Please do not equate value guide price ranges with actual prices. These ranges are suggestions that are subject to variations and change, and *neither the authors nor the publisher can be responsible for any outcomes from consulting this guide*. We do, however, wish you luck in the hunt and pleasure in owning beads in fashion.

PART II

Chapter 5
Dresses & Gowns

Above: Pink silk chiffon. The gown is hand beaded with pink and white satin finish bugle beads to create the large flowers. Silver-lined rocailles are hand stitched to form the smaller leaves. c.1920s. $300-500. *Courtesy Ursuline College Historic Costume Study Collection*

Top left: Peach silk lace Coco Chanel design knock-off. Hand beaded in dyed peach seed beads that trace the design in the lace. This sleeveless gown has a scoop neckline, dropped waistline, peach satin shoulder bow tie, and peach satin belted bow at waistline. c.1925. $800-1200. *Courtesy Ursuline College Historic Costume Study Collection*

Bottom left: Detail.

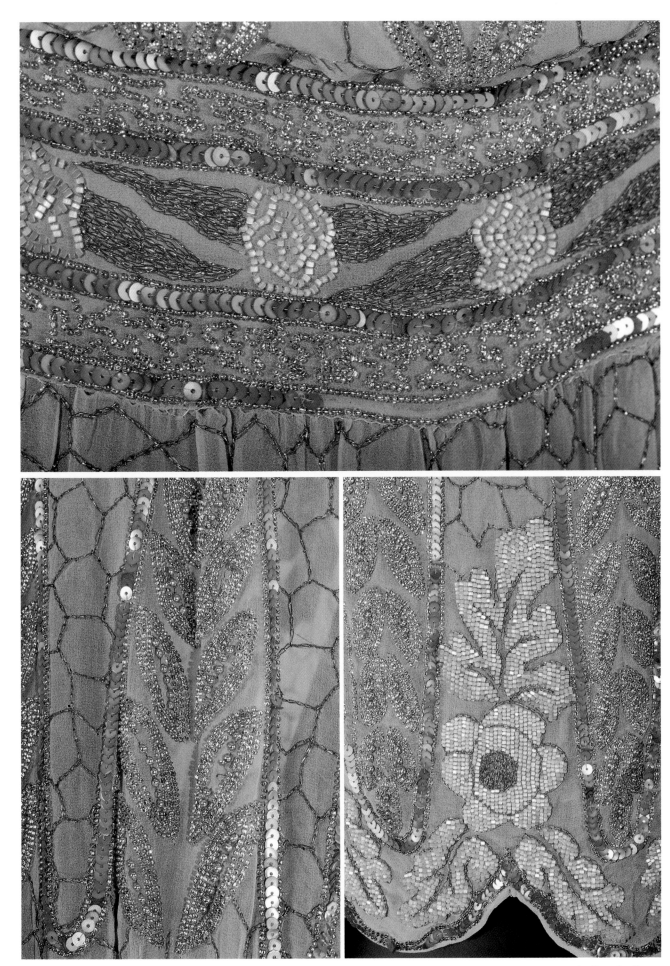

Bottom left: Detail of skirt.

Top: Detail of belt.

Bottom right: Detail of hem.

Rose silk chiffon. The "flapper" dress has a tiered skirt in triangles, slit ribbon bodice and is tambour beaded in satin finish bugle beads with set diamanté trim. c.1920s. $300-400. *Courtesy Estate of Zelta Schulist Glick*

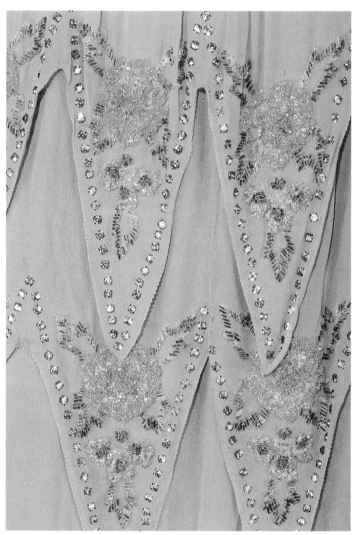

Detail.

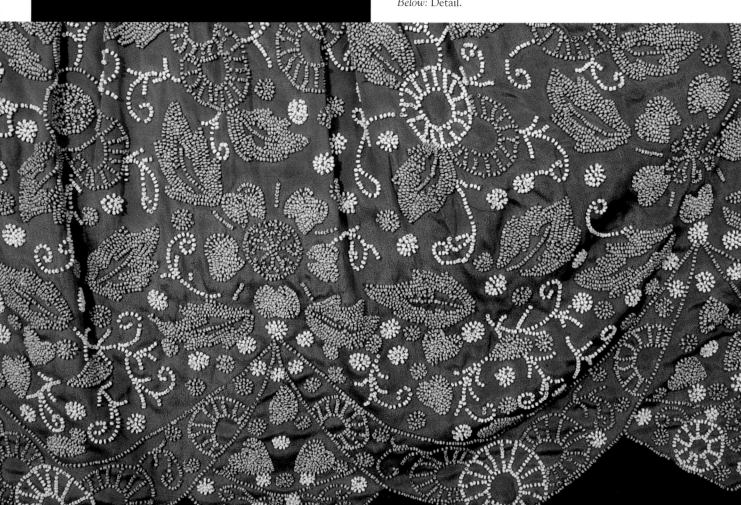

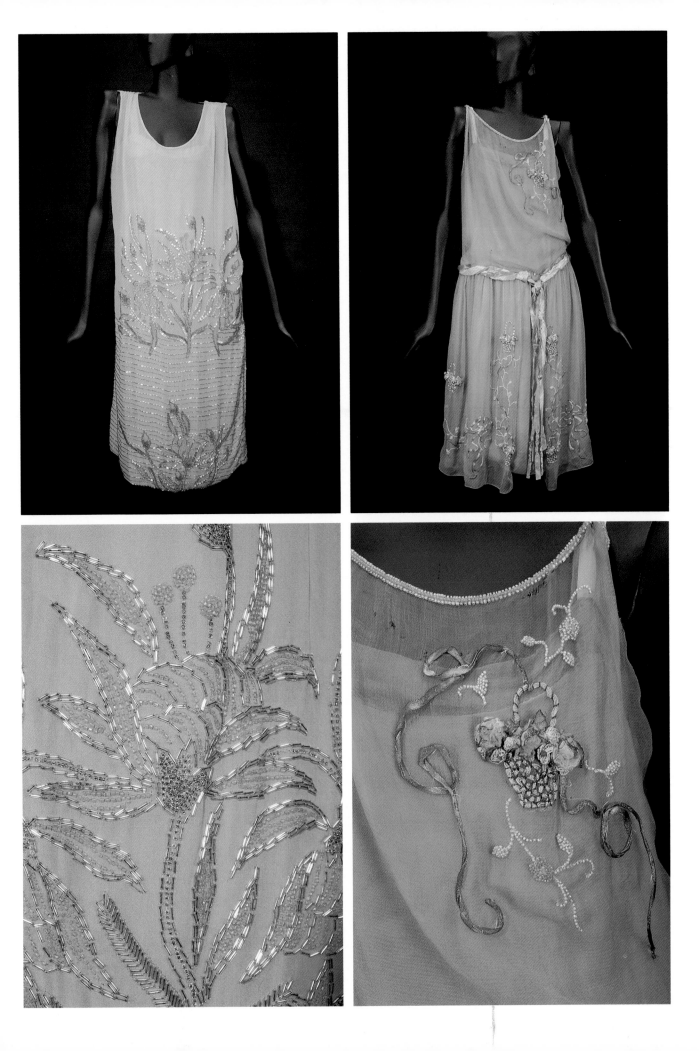

Top left: Aqua silk. Tambour beading in mother-of-pearl bugles on bodice and skirt form wide bands of floral designs and a center medallion. c.1920s. $200-300. *Courtesy Charlotte Michell Trenkamp*

Bottom left: Detail.

Top right: Ivory silk netting over peach silk. The silk netting is hand beaded with gold and silver-lined bugles. Mauve bugles are tambour stitched creating matching patterns to the hand embroidery in silver thread. c.1920s. $150-250. *Courtesy Doreen Leaf-Hund*

Bottom right: Detail.

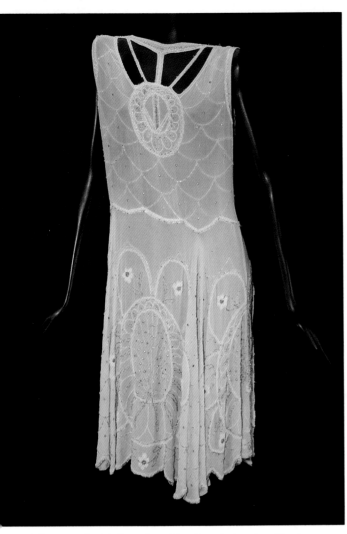

Top left: Ivory silk. Mother-of-pearl beads and silver-lined bugles are hand beaded in Art Deco floral designs and accented with rhinestones. The open cut pattern of the bodice is repeated in beads on the center medallion. c.1927. $150-250. *Courtesy Travis Abbott for Catherine Elizabeth Szabo*

Top right: Black silk chiffon. The Art Deco dress is tambour beaded with white opaque and crystal bugles, set diamanté forming floral designs on the skirt. c.1925. $150-200. *Courtesy Paulette Batt, Larchmere Antiques, & Paulette's Vintage*

Bottom right: Back view.

Above: Black silk net. The silk netting is hand beaded with clear and black sequins and jet bugles in Art Deco patterns. Jet faceted drop beads and fringe embellish the design. c.1920s. $150-250. *Courtesy Doreen Leaf-Hund*

Top left: Black silk chiffon. Tambour stitched with jet beads, the Parisian evening gown features a "bat- wing" attached cape. c.1920s. $150-250. *Courtesy Linda Katz*

Bottom left: Detail.

Above: Mauve silk. The designs on the bodice and skirt are hand beaded with crystal seed beads. c.1930s. $150-250. *Courtesy Sally Smith*

Top right: Teal silk crepe. The bodice, hip drape, and sash are hand beaded in silver-lined bugles in vermicelli and floral patterns. c.1930s. $150-250. *Courtesy Emma Lincoln*

Bottom right: Detail.

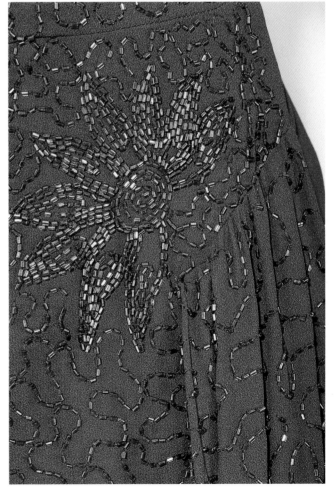

Above: Black silk. Jet bugles and cut beads are tambour beaded in an overall feather design. Rows of jet bugles create a belt and rhinestones outline a beaded bow at the dropped waistline. c.1920s. $200-300. *Courtesy Anna Greenfield*

Top left: Black silk. The hand tambour stitched beading is jet bugles, tiny embossed black sequins, and gold bugle beads forming a "leaf" and swirl design. c.1920s. $200-300. *Courtesy Ursuline College Historic Costume Study Collection*

Bottom left: Detail.

Opposite page:
Top left: Black silk chiffon. The sleeves, bodice, and panels on the skirt are hand beaded in amber colored seed beads and royal blue cut beads in a lattice pattern, c.1930s. $200-300.

Bottom left: Detail.

Top right: Black silk chiffon. Hand tambour beading of jet and royal blue cut beads form Art Deco designs, medallions, circles, and "vermicelli" pattern. c.1920s. $250-300. *Courtesy Janet King Mednik*

Bottom right: Detail.

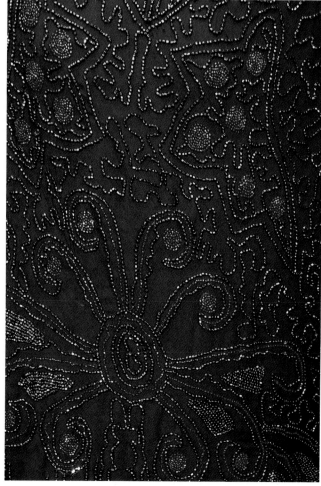

Opposite page:

Top left: Blue silk. Blue iridescent cut beads are hand worked in medallion patterns, in rows, and looped on the edge of the bateau neckline and sleeves. c.1930s. $100-150. *Courtesy Emma Lincoln*

Bottom left: Black silk chiffon. The Art Deco medallion and triangular patterns are hand tambour stitched in jet cut beads. c.1930s. $150-200. *Courtesy Rebecca Smith*

Top right: Black silk chiffon. Hand tambour beading with jet bugles and black faceted beads create an overlapped diamond pattern on the skirt. Gilded thread embroidery embellishes the medallions on the waistline. c.1920s. $250-350.

Bottom right: Detail.

Black silk net and teal satin. The netting is trimmed on the straps and on the skirt with jet fringe of faceted drop beads, bugles, and black sequins. c.1930s. $100-150. *Courtesy Ursuline College Historic Costume Study Collection*

Black silk georgette. Hand beading in white opaque seed beads trims the bateau neck and creates the wide bands of Art Deco designs on the sleeves and on the skirt. White seed beads are hand worked in geometric patterns and a repeated floral motif. c.1930s. $300-400. *Courtesy Anna Greenfield*

Above: Detail of opposite page.

Bottom left: Black crepe. The gown is trimmed with jet beads at the neckline and hand beaded closures on the bodice. c.1930s. $100-150. *Courtesy Ursuline College Historic Costume Study Collection*

Bottom right: Detail.

Top left: Black lace and pink silk. The bodice of the evening gown is tambour stitched in an Art Deco design with satin finish bugle beads and set diamanté. c.1930s. $300-400. *Courtesy Ursuline College Historic Costume Study Collection*

Bottom left: Detail.

Top right: Black silk taffeta. The bodice, sleeves, and skirt are trimmed with fringe of jet bugles and cut beads. Loose strings of cut jet beads are hand beaded on the shoulder. The lace trimmed dicky appears to be an added decoration to lessen the depth of the décolletage, possibly to make the dress suitable for daytime wear. c.1930s. $100-200. *Courtesy Anna Greenfield*

Bottom right: Detail.

Top left: Rust silk crepe. The bodice is hand beaded in bronze metallic cut beads and bugles. Similar beads are used in the necklace and on the bag adorning the mannequin. c.1940s. $75-125. *Courtesy Ursuline College Historic Costume Study Collection*

Bottom left: Pink silk. White pearls and crystal rhinestones accent the tatted ribbon embroidery in floral and leaf patterns on the wide cape-like collar, bodice front, and on the skirt. c.1950s. $100-200. *Courtesy Anna Greenfield*

Top right: Navy rayon crepe. The bodice, sleeves, and waist are hand beaded in swirls of blue iridescent bugle beads. Beaded sphere tassels of blue cut beads are hung at the neck and at the waist. Eisenberg & Sons for The Halle Bros., Cleveland, Ohio. c.1940s. $300-400. *Courtesy Donna Kaminsky*

Bottom right: Detail.

Top left: Pale gray silk strapless evening gown. The scalloped bodice has elaborate hand work in silver embroidered braid, pearls, silver-lined bugles, rhinestones, and iridescent sequins. c.1950s. $100-200. *Courtesy Anna Greenfield*

Bottom left: Detail.

Top right: Champagne silk chiffon. The empire evening gown has a hand beaded yoke covered in a lattice pattern in gold filled seed and bugle beads. The cap sleeves and yoke are encrusted with pearls, plastic faceted beads, gold filled seed and bugle beads, rhinestones and square "frame" gold sequins. Bullocks at Beverly Hills Wilshire. c.1950s. $200-350. *Courtesy Pamela F. LaMantia*

Bottom right: Detail.

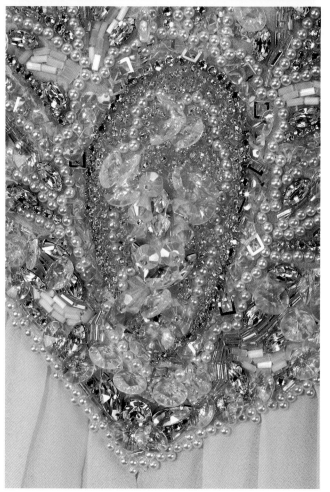

Top left: Blue silk and white lace. The lace neckline and bodice are hand beaded with crystal seed beads strung in fringes and silver-lined bugles and rhinestones. c.1950s. $125-175. *Courtesy Ursuline College Historic Costume Study Collection*

Top right: Green silk net. The bodice and the skirt of the strapless gown have two layers of net over matching silk. The netting is hand beaded and embroidered with gilded cording. The pink and green cut beads and round and pear shaped pearls form a cascade of flowers on the bodice to the waistline. c.1940s. $75-125. *Courtesy Ursuline College Historic Costume Study Collection*

Bottom right: Ivory satin and lace. The evening gown has two layers of lace. The surface netted lace is hand worked in satin flowers that are beaded in pearls, silver-lined bugles, and rhinestones. c.1950s. $100-200. *Courtesy of Ursuline College Historic Costume Study Collection*

Gold lamé, black lace, and chiffon. The heart shaped bodice is of beaded black lace over gold lamé with velvet straps. Hand beaded black faceted beads create the centers of the floral patterns of the lace. c.1950s. $100-150. *Courtesy Ursuline College Historic Costume Study Collection*

Detail.

Gold silk chiffon. The bodice is hand beaded in a paisley pattern with satin finish and amber seed beads. c.1950s. $100-250. *Courtesy Ursuline College Historic Costume Study Collection*

Detail.

Pink nylon net. Netting is machine beaded in all-over design with crystal seed beads and rice shaped pearls to create ribbon swirl patterns. Faux pearl faceted beads form tiny stars on the neckline, cuff, and hemline. c.1960s. $150-250. *Courtesy Anna Greenfield*

Detail.

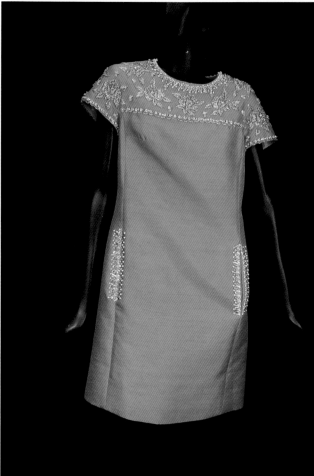

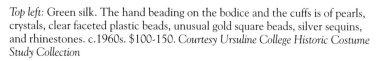

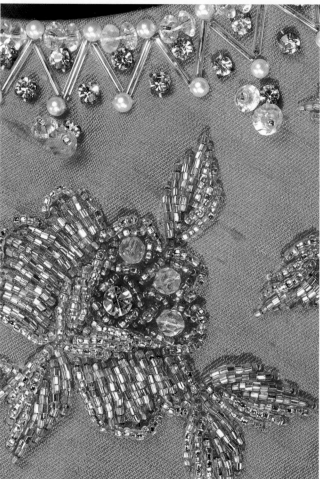

Top left: Green silk. The hand beading on the bodice and the cuffs is of pearls, crystals, clear faceted plastic beads, unusual gold square beads, silver sequins, and rhinestones. c.1960s. $100-150. *Courtesy Ursuline College Historic Costume Study Collection*

Bottom left: Detail.

Top right: Hot pink silk shantung. The trim on yoke, neck, sleeves, and slit pockets is hand beaded with silver-lined seed and bugle beads, pearls, clear faceted plastic beads, and rhinestones. c.1960s. $75-125. *Courtesy Ursuline College Historic Costume Study Collection*

Bottom right: Detail.

Top left: Yellow silk. Trapeze silhouette with a stand up collar is hand beaded and encrusted with rhinestones, silver-lined crystal, and faceted iridescent plastic beads. c.1960s. $75-125. *Courtesy Ursuline College Historic Costume Study Collection*

Bottom left: Detail.

Top right: Blue linen. A-line silhouette with cap sleeves and jewel neckline that are hand beaded and embroidered with jewels of crystal bugles; white, pink, and gold-lined seed beads; pink and green rhinestones; large oval faceted white beads; and metallic threads in pink and silver. c.1960s. $75-125. *Courtesy Wilma Simon Trenkamp*

Bottom right: Detail.

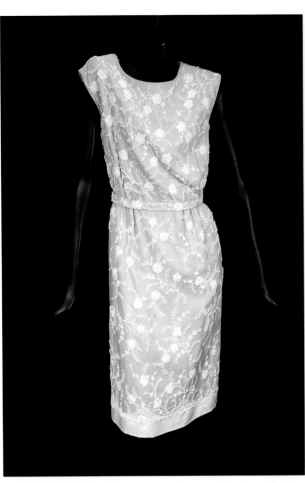

Top left: Pink and white linen. An all-over floral design is machine embroidered and beaded in white opaque seed beads, larger white round beads, and white silk thread. c.1960s. $100-150. *Courtesy Barbara Szabo Galbos*

Top right: Blue and white rayon. The white fabric of the bodice joins the blue in jagged points and is hand beaded in white opaque seed beads. c.1960s. $75-100. *Courtesy Ursuline College Historic Costume Study Collection*

Detail.

Top left: Green silk. The bodice is richly hand beaded with crystal seed beads and bugles, accented with silver sequins in a lattice pattern. A loose looped serpentine pattern of crystal beads decorates the neck and continues in two rows down the front of the dress to the hem. c.1960s. $100-150. *Courtesy Anna Greenfield*

Bottom left: Detail.

Top right: Pink silk. The bodice and skirt of the empire style gown are hand beaded with silver-lined bugles, crystals, and rhinestones. c.1960s. $75-125. *Courtesy Ursuline College Historic Costume Study Collection*

Bottom right: Ivory silk crepe. The bodice of the empire style gown is hand beaded in a "vermicelli" pattern and encrusted with flowers and loops in crystal and gold-lined round and bugle beads. c.1960s. $75-125. *Courtesy Virginia Barry*

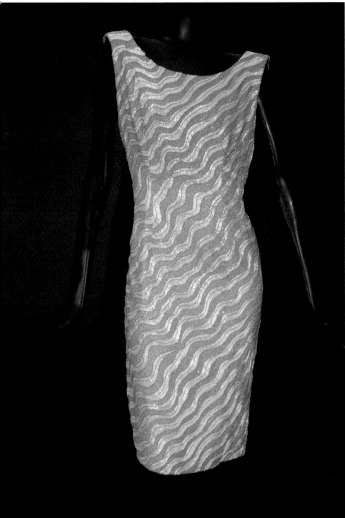

Green silk. The sheath dress with scoop neckline is beaded with green glass faceted beads, tambour stitched with bead rows, and waves of satin finish bugle beads. Guy Laroche designer. c.1960s. $400-500. *Courtesy Ursuline College Historic Costume Study Collection (New acquisition from The Museum of The Fashion Institute of Technology)*

Detail.

Opposite page:
Top left: Red rayon crepe. The band around the empire styled bodice is accented with red rhinestones and faceted acrylic beads. c.1960s. $50-75. *Courtesy Ursuline College Historic Costume Study Collection*

Top right: Green silk crepe. The bodice of the empire gown is hand embroidered and beaded with flower appliqués on net with pink round and rice shaped pearls, pink bugles, iridescent plastic balls, clear faceted plastic drops, pearl drops, rhinestones, and pink and green sequins. c.1960s. $75-125. *Courtesy and handmade by Emma Lincoln*

Bottom left: Black and white silk. The white silk bodice is covered in iridescent sequins and beaded with silver-lined bugles to create a checkerboard pattern, c.1960s. $75-125. *Courtesy Ursuline College Historic Costume Study Collection*

Bottom right: Black silk chiffon. The sweetheart bodice and cap sleeves of the empire gown are hand beaded with faux jet seed, bugle beads, and faceted beads. Liz Claiborne Design, made in China. c.1960s. $75-125. *Courtesy Virginia Mallett Azzolina*

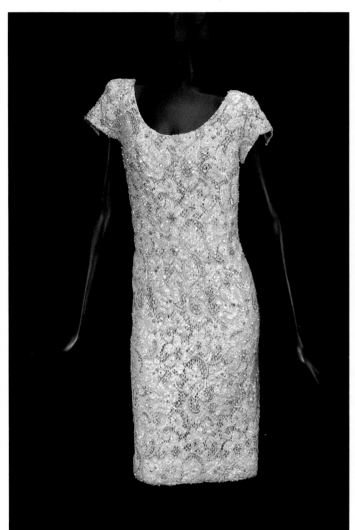

Ivory lace. The crocheted lace fabric is hand embroidered in aurora borealis crystal beads, pearls, iridescent sequins, golden rhinestones, and tiny gold sequins to accent the floral pattern. c.1960s. $300-400.
Courtesy Ursuline College Historic Costume Study Collection

Detail.

Raspberry wool knit. Matching sequins cover the dress. Hand beaded pearls and mother-of-pearl bugles form the flowers, leaves, and vines on the sleeves, bodice, and skirt. Imperial Imports, made in British Crown Colony of Hong Kong. c.1960s. $200-300. *Courtesy Ursuline College Historic Costume Study Collection*

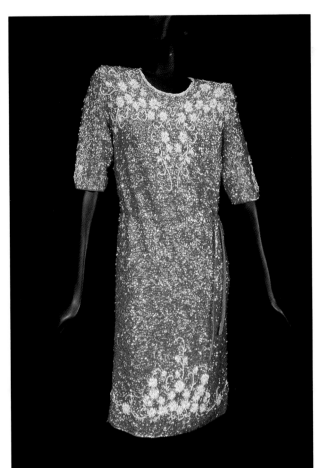

Detail.

Top left: Yellow silk chiffon. The bodice is heavily beaded by hand with white opaque seed beads and white sequins in an all-over "vermicelli" pattern. At intervals, small flowers are formed by white seed beads looped to give added dimension. Karen Stark designer for Harvey Berin. Milgrim label. c.1960s. $150-250. *Courtesy Sarah N. Sato*

Top right: Yellow silk chiffon. The bodice of the empire gown is hand beaded with gold-lined and crystal beads and embroidered with amber and orange silk thread. Milgrim label. c.1960s. $75-125. *Courtesy Sarah N. Sato*

Bottom left: Yellow and black silk with black lace. The re-embroidered lace on black net is accented with faux jet cut and faceted beads that drop as tassels from the center of the lace flowers. c.1960s. $100-200. *Courtesy Rebecca Smith*

Bottom right: Black wool knit. Wide silk fringe trims the bateau neckline. Black faceted beads and tear drops are evenly spaced on the bodice to the hip line. c.1960s. $75-125. *Courtesy Ursuline College Historic Costume Study Collection*

Top left: The rainbow pastel printed silk organza evening dress is hand beaded in gold-lined bugles and gold sequins with gold thread following the printed pattern on the silk. Both the mandarin collar and belt are encrusted with pink iridescent faceted beads, crystals, and gold, pink, and mirrored rhinestones. George Halley - New York. c.1960s. $200-300. *Courtesy Sarah N. Sato*

Bottom left: Detail.

Top right: Green silk paisley. The paisley print trapeze dress has a hand beaded band across the bodice yoke. c.1960s. $150-250. *Courtesy Sarah N. Sato*

Bottom right: Detail.

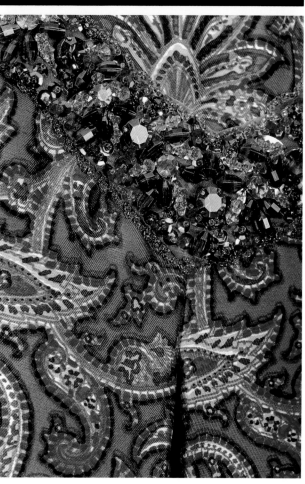

Chartreuse linen. The A-line dress with subtle open weave bodice and center front pleat has a hand beaded bodice using bugle and round white molded beads. Jean Dessès designer. Made for Jean Dessès En Grèce label. c.1967. $250-350. *Courtesy Ursuline College Historic Costume Study Collection (New acquisition from The Museum at The Fashion Institute of Technology)*

Detail.

White and yellow silk print. The sheath cocktail dress is covered with sequins and beads over a yellow floral printed silk. c.1960s. $150-250. *Courtesy Cynthia Barta, Legacy Antiques and Vintage Clothing*

Detail.

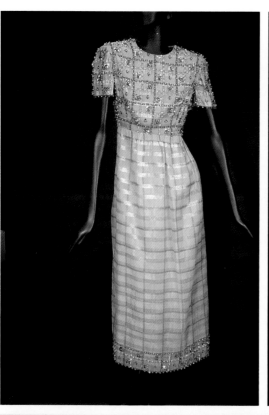 

Top left: Chartreuse, fuchsia, and gilded thread plaid silk organza. The Geoffrey Beene inspired empire gown is hand beaded with chartreuse and fuchsia faceted plastic beads and matching sequins, which trace the plaid design of the fabric on the bodice, sleeves, and hem. Elizabeth Arden label. c.1960s. $150-250. *Courtesy Ursuline College Historic Costume Study Collection*

Top right: Ivory silk. The evening gown has 9 inches of hand beading, resembling an Egyptian collar, covering the bodice and cap sleeves. The bead embroidery is worked on net in white and iridescent silver-lined bugles and pink and turquoise rhinestones. c.1960s. $250-350. *Courtesy Ursuline College Historic Costume Study Collection*

Oppostie page:
Detail of top right photo.

Below: Detail of top left photo.

Top left: Green silk. The bodice of the sheath gown has an attached camisole in an overlapping petal pattern traced with silver-lined beads and iridescent sequins. c.1970s. $75-100. *Courtesy Ursuline College Historic Costume Study Collection*

Top right: Blue satin. The sheath gown has tiny black opaque seed beads hand worked in a floral pattern on bodice and skirt. A black chiffon scarf is attached at the shoulder. c.1970s. $75-125. *Courtesy Ursuline College Historic Costume Study Collection*

Bottom left: Pink silk. The evening gown with tambour stitched pink iridescent sequins, forming the background for "paisley" shapes created by pink pearls and seed beads. c.1970s. $150-250. *Courtesy Estate of Marisa Birnbaum*

Bottom right: Detail.

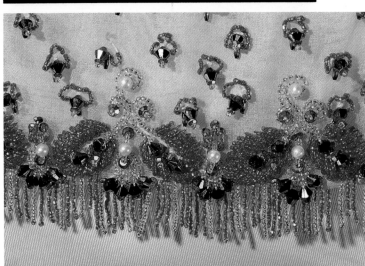

Top left: Pink silk. The organza bodice is hand embroidered with satin finish and crystal bugle beads, satin thread, and roses. The 2-inch fringe trimming the bodice is in crystal bugles and accented with large faceted crystal round beads. Cristobal Balenciaga designer, Spanish. c.1970s. $600-800. *Courtesy Ursuline College Historic Costume Study Collection (New acquisition from The Museum at The Fashion Institution of Technology)*

Bottom left: Detail.

Top right: Cream silk. The bodice of the gown is hand beaded in gold-lined and red seed beads. The design is accented with charcoal and gold bi-cone crystals and pearls. c.1970s. $200-300. *Courtesy Ursuline College Historic Costume Study Collection (New acquisition from The Museum at The Fashion Institute of Technology)*

Bottom right: Detail.

Gold silk taffeta. The long evening gown with
dramatic bustle back is hand beaded on the bodice
and flounce sleeves with gold-lined rocailles.
c.1970s. $200-300. *Courtesy Ursuline College
Historic Costume Study Collection (New acquisition
from The Museum at The Fashion Institute of
Technology)*

Detail.

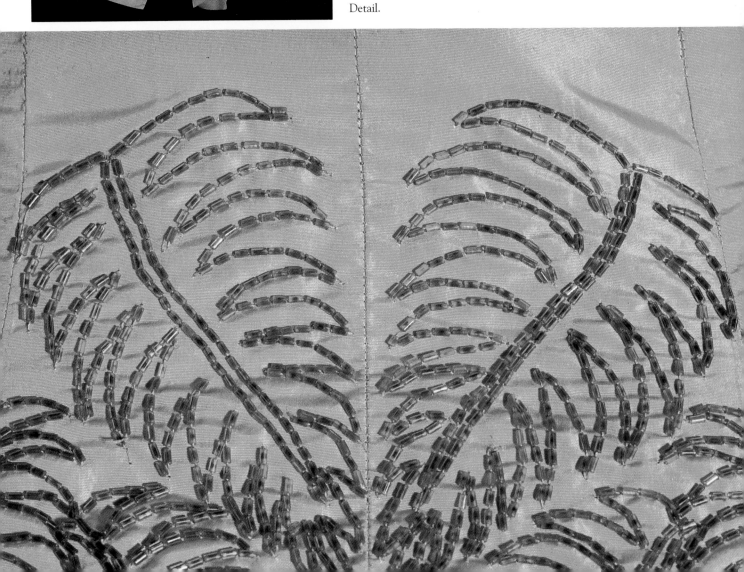

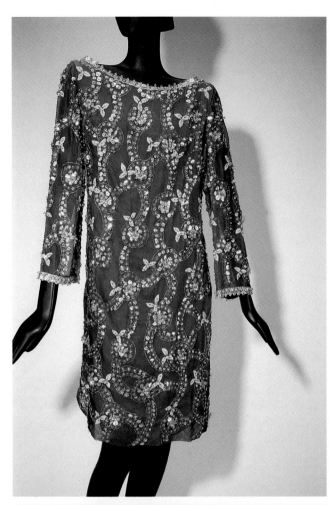

Above: Green silk. The "quilt stitch" pattern is embroidered with gilded thread and each diamond is accented with gold drop beads. Caché, made in Korea. c.1970s. $100-150. *Courtesy Houri Askari*

Top left: Olive green silk chiffon. The sheath dress is hand beaded in swirl and circle designs with gold-lined round and bugle beads, and gold sequins in a variety of shapes. Leaf shape gold sequins accent the flower designs in the circles. Malcolm Starr designer. c.1980s. $150-200. *Courtesy Ursuline College Historic Costume Study Collection*

Bottom left: Detail.

Opposite page:
Top left: Pink chiffon. The gown is tambour beaded in a leaf pattern with sequins and silver-lined bugle beads. Laura Salkin Bridals & Fashions. c.1980s. $100-200. *Courtesy Estate of Zelta Schulist Glick*

Bottom left: Pink silk organza. The long sleeve gown is hand embroidered with silver bugles, large silver sequins, and accented with faceted plastic beads and rhinestones. Malcolm Starr designer. c.1980s. $150-200. *Courtesy Ursuline College Historic Costume Study Collection*

Top right: Yellow silk. The organza bodice is hand beaded with pearls, crystal, silver-lined round and bugle beads, faceted crystal drops and rhinestones. Hathaway label, Shaker Heights, Ohio. Made in British Crown Colony of Hong Kong. c.1980s. $250-300. *Courtesy Emma Lincoln*

Bottom right: Detail.

Above: Lavender blue silk jacquard. Hand beaded trims on the collar and cuffs and the abstract designs on the dress are created with lavender seed beads, iridescent sequins, and bugle beads of white satin finish, silver-lined, and purple. A 4-inch fringe of lavender seed beads hangs from the end of the self belt. c.1980s. $100-150. *Courtesy Ursuline College Historic Costume Study Collection*

Top left: Royal blue silk chiffon. The cocktail ensemble has floral designs tambour stitched in silver-lined round and bugle beads and accented with rhinestones. c.1970s. $100-150. *Courtesy Emma Lincoln*

Bottom left: Royal blue polyester chiffon. The bodice is covered in matching net that is beaded with royal blue seed beads, silver-lined blue bugles, and iridescent sequins. Jack Byron label. c.1980s. $100-150. *Courtesy Lillian Birnbaum Horvat*

Opposite page:
Two details of lavender blue silk jacquard.

Above: White silk lace. The gown has hand beaded accents on the high neck and down the front with rice shaped pearls and iridescent sequins. c.1980s. $200-300. *Courtesy Katherine O'Neill*

Top left: Ivory silk chiffon. The beaded bodice, sleeves, and layered skirt are hand tambour stitched with pearls, mother-of-pearl bugles, and iridescent sequins. Lillie Rubin Exclusive Design, made in China. c.1995. $150-200. *Courtesy Zillah Solonche Green*

Bottom left: Ivory silk chiffon. The tambour beading in an ivy design of pearls, round and bugle crystal beads, and white sequins. Made in India. c.1980s. $100-150. *Courtesy Judith Letteer Baidy*

Opposite page:
Top left: Ivory silk lace. The lace is re-embroidered with satin finish bugle beads, crystal seed beads, and silver sequins. Victoria Royal Ltd., made in British Crown Colony of Hong Kong. c.1986. $200-300. *Courtesy Betty Weber Bolwell*

Bottom left: Detail.

Top right: White silk chiffon. The gown is tambour stitched by machine in an all-over "vermicelli" pattern with crystal bugles. The fans of leaves and vines are designed with silver-lined bugles, silver balls, sequins, and an accent of rhinestones. Black Tie Design. c.1990. $200-300. *Courtesy Lucia Zavarella Fedeli*

Bottom right: Detail.

Above: Detail.

Bottom left: Black silk. Black cut, and bugle and faceted beads are hand worked to create a 6-inch fringe on the shoulders and are embroidered in designs and fringe on the neckline and down the front of the gown. c.1980s. $200-300. *Courtesy Emma Lincoln*

Bottom right: Black silk chiffon. The bodice, designed to resemble a cape with a sweetheart neckline, is beaded in silver-lined seed beads, silver sequins, pearls and satin finish bugles, all of which form the floral design. Laurence Kazar - New York, made in India. c.1980s. $150-200. *Courtesy Mari Stanek Hageman*

Opposite page: Detail.

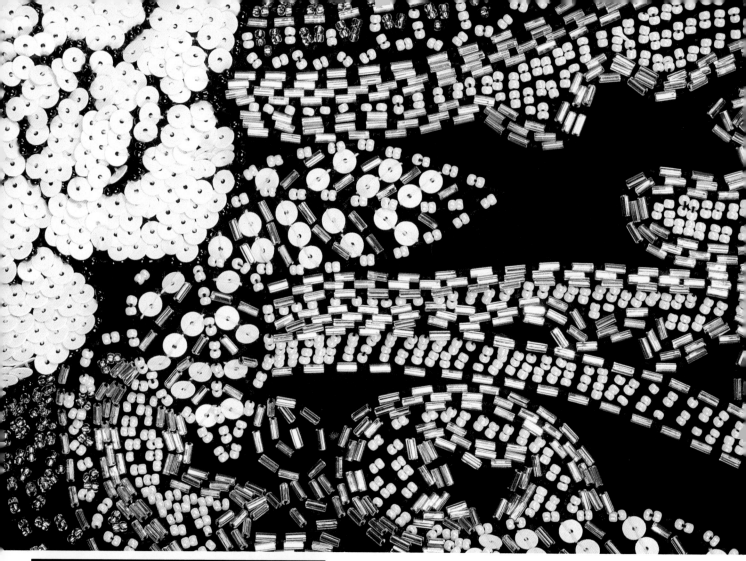

Detail.

Black chiffon. Bodice of the gown is beaded
with white luster and crystal seed beads, silver-
lined bugles, and white sequins in a floral motif.
Black Tie Design. c.1990s. $200-250. *Courtesy
Anna Greenfield*

Opposite page:
Top left: Fuchsia silk chiffon. The dress is tambour stitched by machine
in an all-over "woven basket" pattern of fuchsia bugle beads. c.1980s.
$150-200. *Courtesy Zillah Solonche Green*

Top right: Black chiffon. The blouson bodice is trimmed with two
triangular designs at the shoulder and at the dropped waistline, which
are hand beaded in silver-lined bugle beads. Dominic Rompollo for
Jean Roberts, made in U.S.A. c.1992. $200-300. *Courtesy Zillah
Solonche Green*

Bottom left: Black silk chiffon. Black iridescent sequins and cut beads
are tambour stitched to cover the dress and sash belt with a variety of
designs. c.1980s. $100-150. *Courtesy Emma Lincoln*

Bottom right: Black chiffon. The shoulders and panels of the attached
over-blouse are hand beaded with white satin finish bugles. c.1980s.
$100-150. *Courtesy Zillah Solonche Green*

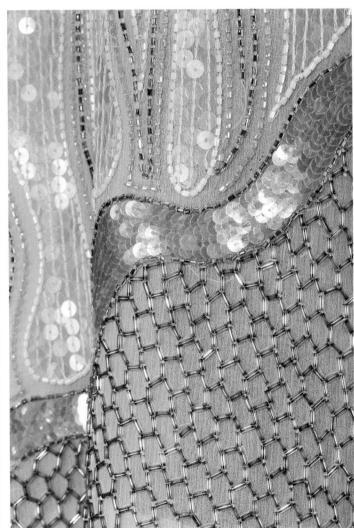

Pink silk chiffon. Iridescent sequins are machine tambour stitched over most of the gown. The "flame" design on the sleeves and the skirt are formed by pink, silver-lined, and mother-of-pearl bugle beads. The hemline has an unusual 10-inch net in a hand beaded off-loom woven pattern of silver-lined bugles bordered with pink iridescent sequins. Made in Montreal, Canada. c.1992. $1000-1500. *Courtesy Nancy C. Goldberg*

Detail.

Cobalt blue silk chiffon. This heavily beaded gown is accented with rhinestones and trimmed with numerous tassels of blue and silver-lined bugles that fall from the bodice, sleeves, and the sides of the skirt. c.1990s. $300-400. *Courtesy Zsuzsa Csepanyi Bawab*

Detail.

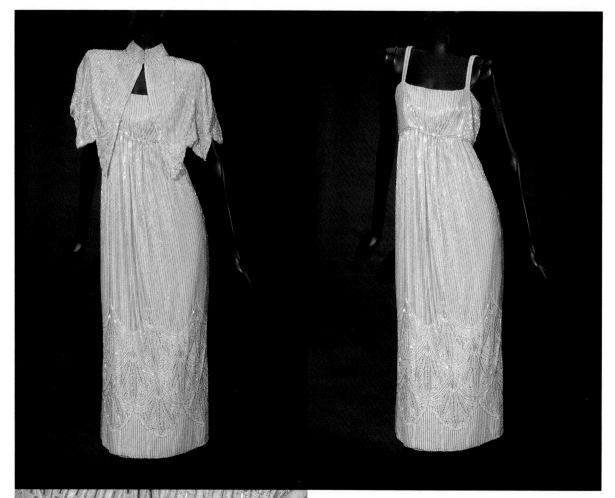

Above: Dress without cape.

Top left: Pink silk lace. The long evening gown and cape are hand tambour beaded with silver-lined bugle beads following the pattern of the lace. c.1990s. $1000-1500. *Courtesy Sarah N. Sato*

Bottom left: Detail.

Detail.

Red wool knit. The sweetheart bodice is hand beaded with pearls, red and blue seed beads, gold-lined seed and bugle beads, and red and blue faceted gems. The cuffs have a trim of pearls and gold-lined, red and blue seed beads. c.1980s. $150-250. *Courtesy Ann D. Jackson*

Three Fortuny gowns in wine, red, and black arranged to display the Murano glass beads attached to the sleeves. Fortuny designs with beads on the sleeves and hems of gowns to provide weight to the hand pleated silk, assuring that the silk falls and drapes elegantly. Mariano Fortuny designer. c.1980s-90s. *Courtesy Phyllis Seltzer*

Black pleated silk. Venetian black luster beads are fastened to both sides of the opening of the coat, hem, sleeves of the coat, and the sides of the gown. Mariano Fortuny designer. c.1990s. $2000-3500. *Courtesy Phyllis Seltzer*

Detail.

Red pleated silk. Venetian striped beads are
attached to the shoulders to weight the drape of
the sleeves, sides, 'V' of the bodice, and hemline of
this two piece gown. Mariano Fortuny designer.
c.1980s. $1500-2500. *Courtesy Phyllis Seltzer*

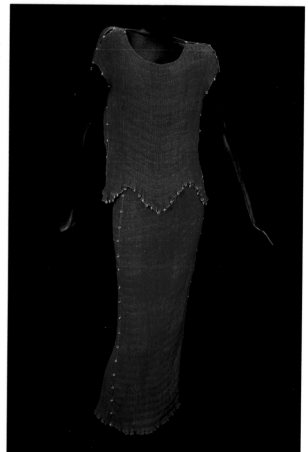

Detail.

Wine pleated silk. Blue iridescent Venetian beads are attached to the sides of the long sleeves and to the hemline to weight the design of the gown. Mariano Fortuny designer. c.1980s. $1500-2500.
Courtesy Phyllis Seltzer

Above: Aqua hand pleated silk. The evening gown with neck and side seams hand beaded with faceted crystals. Russell Trusso designer, Cleveland, Ohio. c.1980s. $500-600. *Courtesy Robin Herrington-Bowen*

Top right: Garnet pleated silk. The bodice of this gown is hand embroidered with glass seed beads the color of garnets, gold-lined round with bugle beads in several measures, gilded thread, rhinestones, and sequins. Mary McFadden designer. c.1980s. $2000-3000. *Courtesy Emma Lincoln*

Bottom right: Detail.

Pink silk crepe. The attached cape is hand painted and hand beaded in silver-lined bugles with 9 inches of cascading fringe in silver-lined bugles. Bob Mackie designer. c.1980s. $3000-4000. *Courtesy Sarah N. Sato*

Back view of cape.

Opposite page: Detail.

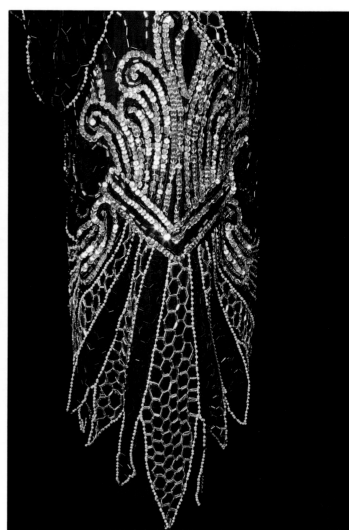

Black silk. The dress is tambour beaded by machine with black and silver-lined bugles, pearls, and silver sequins in a variety of patterns. A large leaf design is beaded in hexagon shapes to fill the leaf forms. c.1990s. $200-300. *Courtesy Estate of Marisa Birnbaum*

Detail.

Black polyester chiffon. The dress is machine tambour stitched in a variety of patterns, nesting circles, veins, and leaves, using white opaque seed beads and crystal bugles, and black iridescent seed beads and bugles. c.1990s. $250-300. *Courtesy Estate of Marisa Birnbaum*

Detail.

Above: Detail.

Bottom left: Black chiffon. The all-over design of leaves is created with white opaque bugles, black seed beads, and black sequins. c.1990s. $150-200. *Courtesy Estate of Marisa Birnbaum*

Bottom right: Gold lamé. The beaded sleeves and low waist bodice are hand tambour stitched with silver-lined round bugle beads, black bugles, gold-lined beads, gold balls with rhinestones, and gold and copper sequins. c.1990s. $250-300. *Courtesy Marilyn Ruckman*

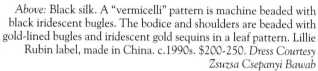

Above: Black silk. A "vermicelli" pattern is machine beaded with black iridescent bugles. The bodice and shoulders are beaded with gold-lined bugles and iridescent gold sequins in a leaf pattern. Lillie Rubin label, made in China. c.1990s. $200-250. *Dress Courtesy Zsuzsa Csepanyi Bawab*

Top right: Royal blue silk with black "eyelash" lace. The tiers of lace on the skirt are hand beaded with 3-inch fringe using black faceted beads and drops in a variety of shapes. James Galanos designer. c.1980s. $1000-1500. *Courtesy Sarah N. Sato*

Bottom right: Detail.

Red wool crepe. The black net insert in a leaf design is hand tambour beaded in gold-lined bugle beads and faceted crystals in a bi-cone shape. Bob Mackie designer. c.1984. $2000-2500. *Courtesy Betty Weber Bolwell*

Detail.

Black silk chiffon. Iridescent black bugles are machine beaded in large floral stems on the bodice. The sleeves and the skirt are tambour stitched in iridescent black sequins and bugle beads. Made in India. c.1990s. $100-150. *Courtesy Lillian Birnbaum Horvat*

Gold silk chiffon. The gold-lined bugles are machine tambour stitched in vertical rows on the upper bodice and skirt. The collar and most of the bodice are hand beaded in floral designs with gold and silver-lined bugles, gold faceted plastic bi-cone beads, and gold sequins. Lillie Rubin, label. c.1990s. $300-400. *Courtesy Lauren Humphrey*

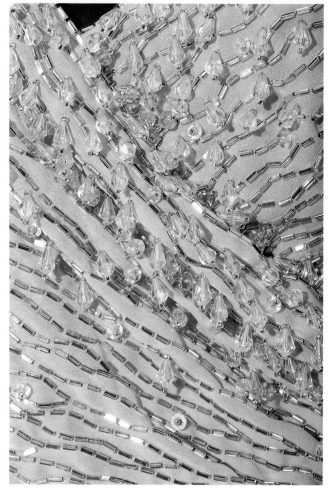

Above: Black chiffon. The Art Deco influenced swirls are machine tambour stitched and beaded in white opaque beads. The use of white thread emphasizes the pattern. c.1995. $400-500. *Courtesy Zillah Solonche Green*

Top left: Blue silk. The long evening gown is beaded by machine and by hand with silver-lined seed and bugle beads accented with faceted crystal drops. Victoria Royal Ltd., made in British Crown Colony of Hong Kong. c.1987. $1500-2000 retail. *Courtesy Betty Weber Bolwell*

Bottom left: Detail.

Pink silk organza. The lace on the bodice is hand embroidered with gold thread and layered over silk that is dotted with gold and sliver glitter. The skirt is hand beaded with light and dark gold bugle beads. Pat Kerr designer. One Cumberland Place, London, England. c.1980s. $12,000 retail. *Courtesy Betty Weber Bolwell*

Detail.

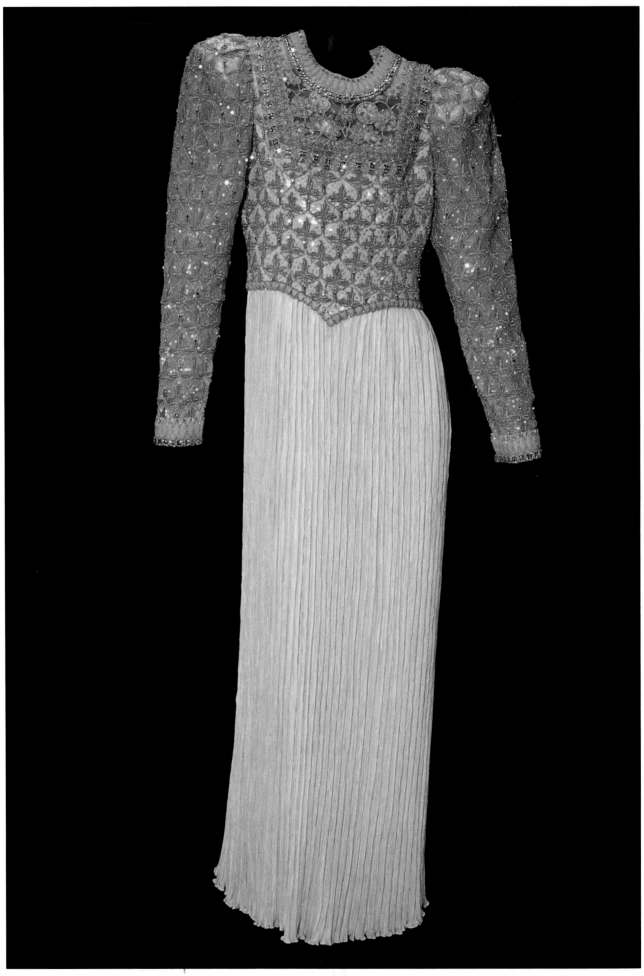

Detail.

Detail.

Detail.

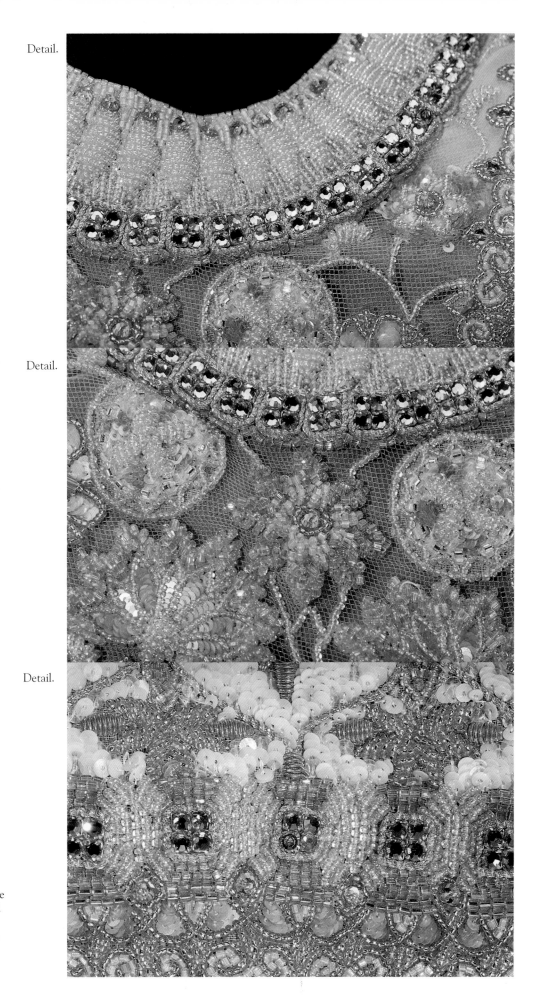

Opposite page:
Cream Fortuny pleated silk. The bodice
and long sleeves are hand embroidered
in pastel seed beads, sequins, and
rhinestones. Mary McFadden designer.
c.1985. $6000 retail. *Courtesy Betty
Weber Bolwell*

Peach silk. This gown is hand beaded in crystal bugle beads, bi-cone   Back view.
faceted crystals, and silver rocailles. The drapes of beaded fringe
cascading from the skirt are 12 inches in length. Bob Mackie designer.
c.1985. $10,000 retail. *Courtesy Carole Carr*

Detail.   Detail.

Top left: From the beginning of the 20th century . . . black silk net. This "illusion" dress is hand embroidered with silk thread and tambour beaded by hand in black faux jet beads. c.1920s. $200-250. *Courtesy Linda Katz*

Bottom left: Detail.

Top right: To the end of the 20th century . . . black micro net. Fashion continues to repeat itself as in this "slipdress," hand and machine beaded in black seed beads. Krizia designer. c.1998. $2000 retail. *Courtesy Nordstrom, Beachwood Place, Cleveland, Ohio*

Bottom right: Detail.

Chapter 6
Separates & Ensembles

Top left: Detail of back of blue denim cotton jacket with hand beaded yoke and back panel in green, red, silver, gold bugle beads, gemstones, rhinestones, and sequins. I. B. Diffusion label. c.1980s. $50-75. *Courtesy Arlene Schreiber*

Top right: Detail of shoulder.

Bottom left: White denim jacket with hand beaded front and back yoke in blue, green, red, silver, gold bugle beads, gemstones, rhinestones, and sequins. I. B. Diffusion label. c.1990. $50-75. *Courtesy Shirley Friedland*

Bottom right: Detail.

Top left: Blue cotton denim jacket with six hand beaded symbols on the front panels, c.1980s. $35-50. *Courtesy Arlene Schreiber*

Bottom left: Detail.

Top right: Blue silk blouse with Southwestern American Indian motif in hand applied bead work on the bodice, c.1990s. $50-75. *Courtesy Arlene Schreiber*

Bottom right: Detail.

Top left: White denim jacket with American Indian motif on the front yoke in coral, turquoise colored beads with silver studs. I. B. Diffusion label. c.1980s. $25-35. *Courtesy Arlene Schreiber*

Top right: Detail.

Bottom left: White cotton blouse with hand beaded collar in American Indian motif, red, white, blue, turquoise, and yellow seed beads, c.1980s. $45-55.

Bottom right: Detail.

Top left: Red evening top, machine tambour stitched with red, blue, and yellow sequins. Yellow seed beads outline huge floral bursts, iridescent bugles, which form the leaves and frame the v-neckline. c.1980s. $150-200 *Courtesy Shirley Friedland*

Top right: Back view.

Below: Detail.

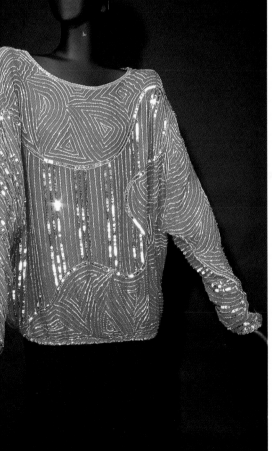

Top left: Black organza "butterfly" blouse, machine tambour beaded in crystal bugles, and silver and royal blue sequins. Made in Germany. c.1980s. $60-70. *Courtesy Krystyna Bryjak*

Top right: Black silk chiffon blouse tambour stitched with black, violet, and blue sequins forming flowers that are dotted with crystal bugle beads. Judith Ann Creations, made in India. c.1980s. $100-125. *Courtesy Houri Askari*

Bottom left: Hot pink chiffon evening top, long sleeves, scoop neckline, machine tambour beaded in crystal bugles, and silver sequins using pink thread that is tambour stitched with silver sequins. c.1980s. $75-100. *Courtesy Janet King Mednik*

Bottom right: Black and gold stripped chiffon blouse with machine tambour beaded gold-lined bugles that outline the stripes, c.1970s. $75-100. *Courtesy Emma Lincoln*

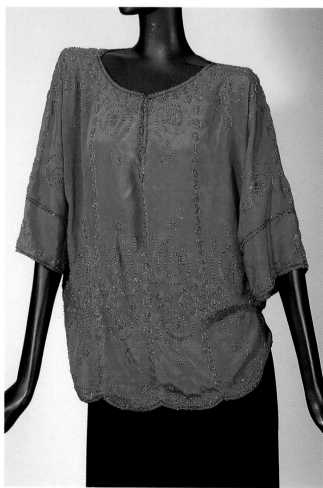

Above: Rust silk blouse, hand beaded with silver-lined rocailles, c.1930s. $35-45. *Courtesy Linda Katz*

Top left: Rust rayon crepe blouse, machine tambour beaded with bronze and gold cut beads in "vermicelli" pattern, c.1950s. $50-60. *Courtesy Anna Greenfield*

Bottom left: Ivory silk chiffon evening top and skirt ensemble. The evening top is hand tambour beaded with gold and satin finish bugles. The sleeve and hem are outlined in gold bugle beads. Laurence Kazar, made in India. c.1996. $250-300. *Courtesy Jerri Lewis Dennis*

Above: Black polyester crepe blouson blouse, hand tambour beaded with black bugles and sequins to create a floral rose at the collar line, c.1980s. $50-75. *Courtesy Anna Greenfield*

Top right: Detail.

Bottom right: Black silk with hand tambour beaded "vermicelli" pattern and a leaf design on sleeve, yoke, and shoulder. Laurence Kazar, made in India. c.1980s. $50-75. *Courtesy Deborah Johnson Rogers*

Top left: Black damask vest hand beaded with black beads and sequins accenting the pattern. Black beaded evening bag with black and red bugles. c.1990s. Vest: $35-40.

Top right: Black chiffon wrap with elbow length sleeves, machine tambour beaded with black iridescent bugles, c.1980s. $50-75.

Bottom left: Black polyester chiffon, layered with hand beaded lace in black glass bugles, which trace floral pattern on lace, c.1980s. $50-75. *Courtesy Emma Lincoln*

Bottom right: Black chiffon long sleeve wrap with machine tambour beading in black iridescent bugle beads and sequins, c.1980s. $50-75.

Top left: Black chiffon long sleeve jacket wrap with hand tambour stitched beading in black bugles and rocailles. Evening bag is beaded in a tulip motif with multi-colored beads. c.1980s. Jacket: $70-80. *Jacket Courtesy Susan Lehman Ellick, Evening bag Courtesy Gloria Azzolina Lorenzo*

Bottom left: Detail.

Top right: White silk crepe tunic with hand beaded sleeves and hem using gold and silver-lined bugles, in a geometric motif, c.1930s. $75-100. *Courtesy Ursuline College Historic Costume Study Collection*

Bottom right: Detail.

Black crepe vest with checkerboard pattern created with hand beaded
rows of gold and crystal rocailles with couched stitching and black
sequins dotted with pearls, c.1970s. $75-100.

Detail.

Turquoise sequined and hand beaded vest with peacock motif outlined with gold balls and multi-colored seed beads, yellow-gold rhinestones, and silver drop beaded fringe, c.1970s. $100-150. *Courtesy Donna Kaminsky*

Detail.

Top left: Green wool knit shell hand embroidered with sequins. Flowers formed by sequins that have pearl centers and the leaves are embroidered with green seed beads. Made in Hong Kong. c.1980s. $100-150. *Courtesy Ethel Dindia*

Bottom left: Detail.

Top right: Black silk shell hand embroidered with black sequins forming a petal design outlined with black bugle beads. Bonwit Teller label. Made in British Crown Colony of Hong Kong. c.1980s. $100-125. *Courtesy Emma Lincoln*

Bottom right: Teal crepe sleeveless shell, hand beaded with crystal seed beads, pearls, and teal sequins. Made in Hong Kong. c.1970s. $75-100.

Top left: White wool knit sleeveless shell, jewel neckline, hand embroidered with iridescent sequins, crystal seed beads, and pearls creating a "V" pattern with fringe. Made in Hong Kong. c.1960s. $75-100. *Courtesy Ursuline College Historic Costume Study Collection*

Bottom left: White wool knit sleeveless shell, scoop neckline, hand embroidered with iridescent sequins, and beaded in a "V" design of fringe in crystal seed beads. Made in Hong Kong. c.1970s. $75-100. *Courtesy Ursuline College Historic Costume Study Collection*

Top right: White wool knit sleeveless shell, scoop neckline, hand embroidered with iridescent sequins, crystal seed beads, and pearls creating a "V" pattern in fringe. Made in Hong Kong. c.1960s. $75-100. *Courtesy Ursuline College Historic Costume Study Collection*

Bottom right: Detail.

Top left: White wool and angora knit sleeveless shell, scoop neckline, hand embroidered with iridescent sequins, pearls, and white luster seed beads. Bottom 3-inch edge is beaded in triangle pattern with pearl centers. The fringe is white luster seed beads. Made in Hong Kong. c.1970s. $100-150.

Bottom left: Blue wool knit sleeveless shell, hand embroidered with iridescent sequins, and beaded with a zigzag pattern and fringe created with crystal bugles and pearls, c.1960s. $75-100.

Top right: White silk shell hand tambour beaded with silver-lined seed beads to form leaves and vines. Dyed pink seed beads form flowers. c.1960s. $50-75. *Courtesy Cynthia Barta of Legacy Antiques and Vintage Clothing*

Bottom right: Detail.

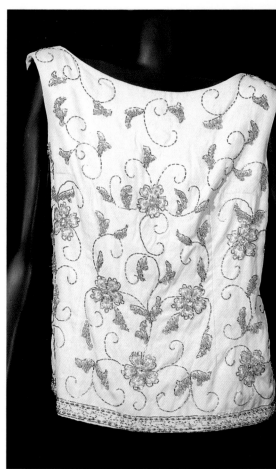

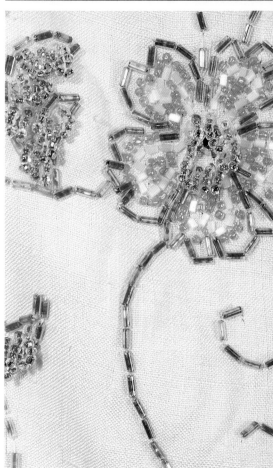

Top left: White cashmere wool knit cardigan with hand beaded floral and leaf motif, in pearl finish seed beads and accented with larger pearls, c.1950s. $125-150. *Courtesy Ursuline College Historic Costume Study Collection*

Top center: Detail.

Bottom left: Black wool knit bolero sweater, hand beaded with crystal bugles, cut crystal beads, and pearls, c.1950s. $40-50. *Courtesy Ursuline College Historic Costume Study Collection*

Top right: Black cashmere cardigan with hand beaded front panels in black bugles, and round and seed beads in floral motif. Black pearl buttons. Kung Brothers, Golden Crown Court, Kowloon & H. K. Hilton Arcade label. c.1960s. $125-150. *Courtesy Paula Ockner*

Bottom right: Detail.

Above: Ivory wool knit cardigan with round collar, dolman sleeve, hand embroidered with iridescent sequins and crystal seed beads, c.1950s. $150-$175. *Courtesy Shirley Friedland*

Top right: White wool knit cashmere cardigan hand beaded with white luster seed beads, gold balls, and iridescent sequins. Pearl buttons. Kung Brothers, Golden Crown Court, Kowloon & H. K. Hilton Arcade label. c.1960s. $125-150. *Courtesy Paula Ockner*

Bottom right: Beige wool knit cardigan trimmed with colored faux shells and faux faceted turquoise square beads, bronze sequins, and bronze cut beads. Pearl buttons. c.1970s. $60-80. *Courtesy Ursuline College Historic Costume Study Collection*

Opposite page:
Top left: Ivory cashmere knit cardigan, hand beaded with bronze cut beads, seed beads, and gold balls. Made in Hong Kong. c.1950s. $150-175. *Courtesy Ursuline College Historic Costume Study Collection*

Bottom left: Detail.

Top right: Black cashmere knit cardigan hand beaded in gold-lined rocailles in a floral motif, c.1950s. $100-125.

Bottom right: Detail.

Top left: Royal blue wool knit cardigan hand embroidered with white seed beads, pearls in floral motif, c.1962. $125-150. *Courtesy Barbara Szabo Galbos*

Top right: Black wool cardigan sweater with turquoise seed beads and "scroll" pattern around all the edges, c.1960s. $125-150. *Courtesy Ruth G. Kyman for Estate of Zelta Schulist Glick*

Bottom: Detail.

Top left: Ivory knit cashmere cardigan, hand beaded with sequins and seed beads over a hand painted floral motif. Made In Hong Kong. c.1960s. $150-200. *Courtesy Ursuline College Historic Costume Study Collection*

Bottom: Detail.

Top right: Long black cardigan with hand beaded pocket watch motif, in silver, gold, and red seed beads, c.1980s. $50-75. *Courtesy Lillian Birnbaum Horvat*

Black ramie and cotton knit cardigan with Op Art design in multi-colored hand embroidery and bead work in glass, plastic, and wooden beads and buttons. The buttons of the sweater are also hand beaded with blue, green, and black beads on white cotton knit. Michael Simon, Henri Bendel, New York label. c. late 1980s. $100-150. *Courtesy Arlene Schreiber*

Detail.

Black ramie and cotton knit cardigan hand worked with multi-colored beads, pearls, buttons, and embroidery, creating cowboy boot design. The buttons are also hand beaded with gold bugles and red rocailles. Michael Simon, Henri Bendel, New York label. c.1990. $100-150. *Courtesy Arlene Schreiber*

Detail.

Top left: Natural colored knitted and crocheted sweater with geometric designs. Bone beads accent the design, and mixed carved bone and unusual glass beads create the fringe. c.1980s. $75-100. *Courtesy Shirley Friedland*

Top right: Black cotton knit sweater hand embroidered with multi-colored wooden and beads, sewn in place with gold rocailles. Emanuel Ungaro designer, made in Hong Kong. c.1980s. $75-100. *Courtesy Katherine O'Neill*

Bottom left: Red wool knit sweater hand beaded with charcoal colored bugles and seed beads in a floral motif, c.1980s. $50-75. *Courtesy Cindy Pressler*

Bottom right: White angora sweater with hand beaded design in charcoal colored bugles, gold rocailles, and bronze sequins, c.1980s. $50-75.

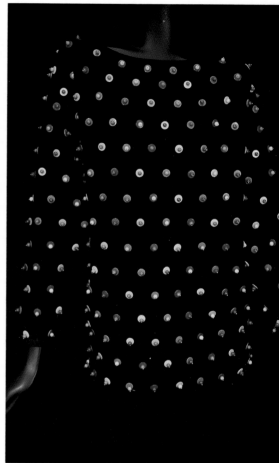

138

Top left: Black pullover beaded sweater with irregular geometric shapes using seed beads, bugle beads in shades of black, and gun metal gray sequins, c.1970s. $75-100. *Courtesy Ruth G. Kyman for Estate of Zelta Schulist Glick*

Top right: Black rayon knit scoop neckline sweater hand beaded with black and white bugle beads in triangle patterns trimmed with camel, royal blue, and white mohair yarn. J. D. Boutique, New York label. c.1985. $75-100

Bottom: Detail.

Top left: Black knit sweater hand embroidered with gold-lined bugles, seed beads and red seed beads, gold balls, pearls, and faceted glass jewels. Sémplice, made in Hong Kong. c.1990. $100-125. *Courtesy Donna Kaminsky*

Bottom: Detail.

Top right: Black wool knit cardigan, band collar, hand beaded with black faceted beads and sequins in a diamond pattern, c.1960s. $75-100. *Courtesy Ursuline College Historic Costume Study Collection*

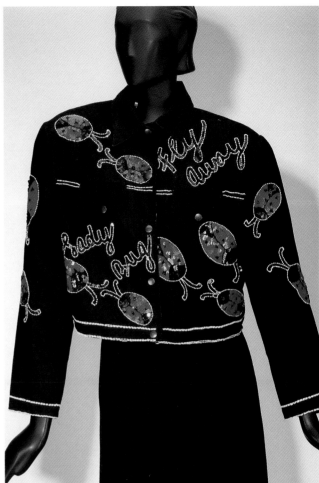

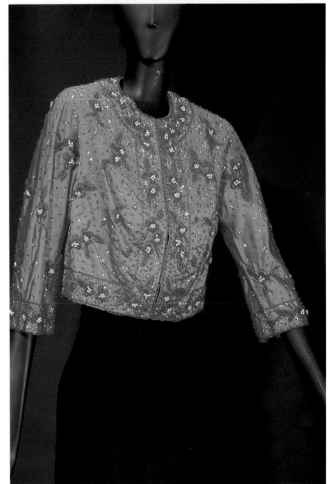

Above: Black chiffon lined long sleeve jacket with tambour beading in black bugles, seed beads, gold-lined bugles, and sequins in a repeated design, c.1980s. $100-125

Top right: Black denim bomber jacket with hand beaded ladybug theme in red and black sequins and pearl finish bugles, c.1990s. $75-100. *Courtesy Marilyn Ruckman*

Bottom right: Red satin bolero jacket, hand beaded with red bugle beads, matching sequins, and white pearls, c.1960s. $150-200. *Courtesy Emma Lincoln*

141

Gray silk organza jacket is hand embroidered with tiny silver gray
sequins, and with silver-lined and crystal bugles and pearls to create a
floral design and the encrusted trim on the collar, cuffs, jacket front,
and hem. Oscar de la Renta designer. c.1980. $500-600. *Courtesy
Emma Lincoln*

Detail.

Black silk jacket with "bracelet" length sleeves hand embroidered in
black and amber sequins, hand beaded with amber and black seed
beads. The jacket is outlined in jet black beads. A black silk skirt
completes the ensemble. James Galanos designer. c.1990s. $600-800.
Courtesy Sarah N. Sato

Detail.

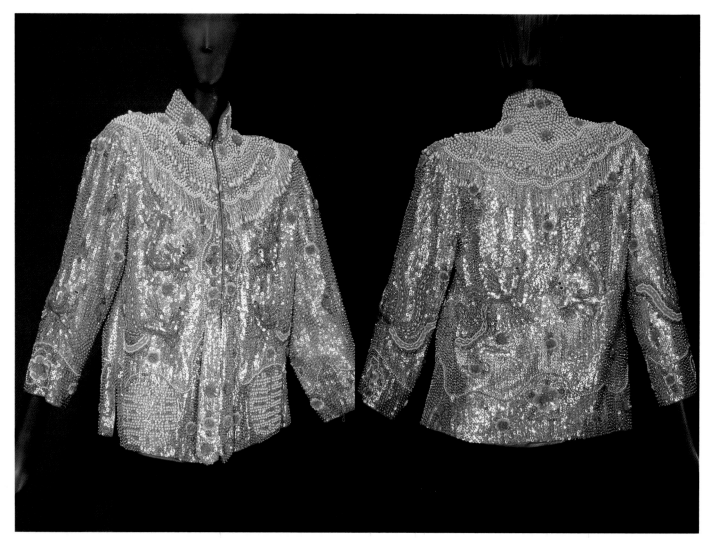

Fuchsia pink silk bead encrusted Chinese jacket lined in red. Dragons are hand embroidered with gold sequins to create raised design. Hand beaded crystal fringe with tear drop pearls circle the yoke. Medallions are covered with pink pearls over gold sequins. c.1960s. $600-800.
Courtesy Emma Lincoln

Back view.

Detail.

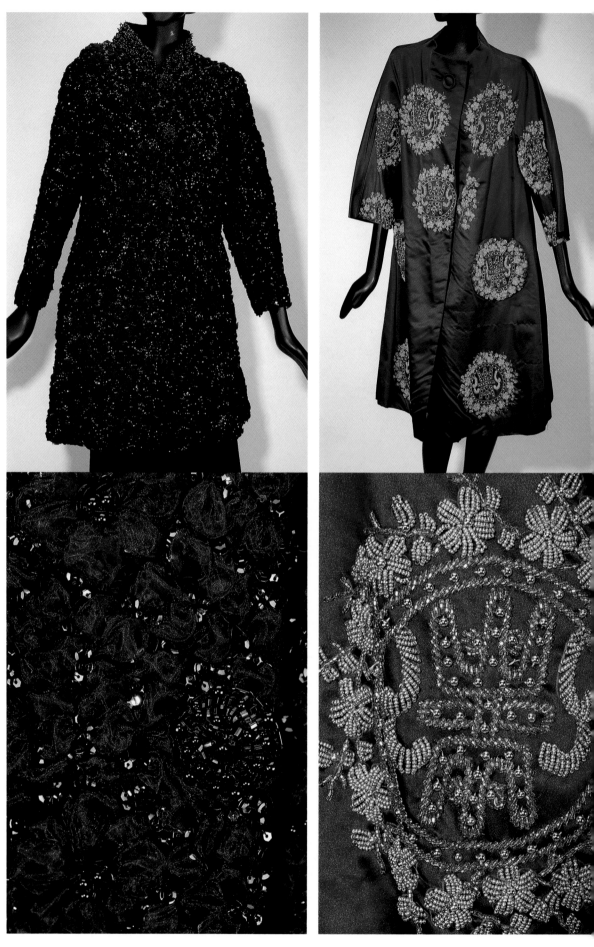

Top left: Black silk coat with silk ribbon forming flowers, accented with black faceted beads. Made in Hong Kong. c.1960s. $200-300. *Courtesy Rebecca Smith*

Bottom left: Detail.

Top right: Emerald green satin wrap coat with large hand beaded Chinese medallions and flowers formed by pale gold rocailles, with gold-filled bugles and gold balls, c.1960s. $200-300. *Courtesy Linda Bowman, Legacy Antiques and Vintage Clothing*

Bottom right: Detail.

Top left: Yellow silk pantsuit, hand beaded on collar and bodice front. Beads are silver-lined, glass bugles, and plastic faceted beads. Rhinestones surround the collar and bodice front. Designed by Nat Allen and custom tailored in Hong Kong. c.1970s. $100-125. *Courtesy Ursuline College Historic Costume Study Collection*

Bottom left: Detail.

Top right: Black wool knit pantsuit with organza sleeve. Pearls, crystal, and black seed beads outline the neck, shoulder, and front tunic top with pearls at the side seam on the trouser. c.1970s. $100-125. *Courtesy Ursuline College Historic Costume Study Collection*

Bottom right: Detail.

Top left: Black wool and rayon knit suit with gold and silver bugle bead chain link rope pattern on cuff, outline of jacket, and hem. Jennifer Roberts, made in U.S.A. c.1980s. $125-150. *Courtesy of Villian Birnhaum Horvat*

Bottom left: Black silk suit. Jacket and skirt are hand tambour beaded with gold and black iridescent bugles, green sequins, and rhinestones. The skirt is tulip shape with "vermicelli" beaded pattern. Fantasi, made in India. c.1990s. $200-300. *Courtesy Houri Askari*

Bottom right: Detail.

Top right: Detail of opposite page photo.

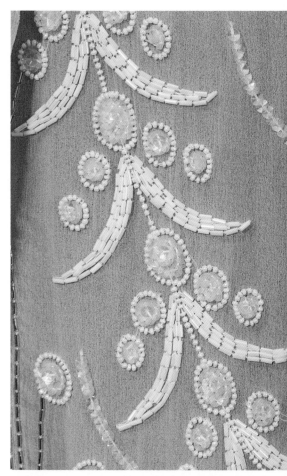

Opposite page: Champagne silk ensemble of dress, scarf, and hat, hand beaded with iridescent sequins, pearls, and silver-lined bugle beads in a variety of patterns. Matching scarf is detachable and can also be worn as a shawl. Judith Ann Creations label. Saks Fifth Avenue. c.1990s. $500-700. *Courtesy Cheryl Byerley Manser*

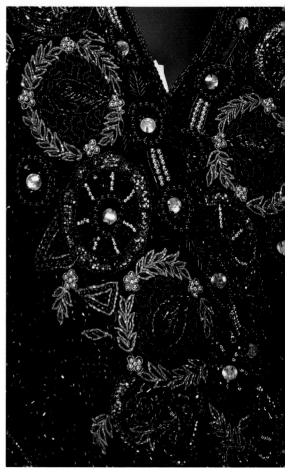

149

Blue silk evenin[g]
coat and gown.
round neck, lon[g]
sleeve, empire w[aist]
coat is hand
embroidered in
silver sequins, g[old]
rocailles, pearls,
and rhinestones
The evening gow[n]
has a scoop
neckline, short
sleeve, empire w[aist]
and is hand
embroidered an[d]
beaded on the
bodice and fron[t]
panel. Christian
Dior design kno[ck]
off. Seaton
Enterprises Ltd.
made in British
Crown Colony [of]
Hong Kong.
c.1980s. $700-9[00]
Courtesy Ursulin[e]
College Historic
Costume Study
Collection

Chapter 7
Accessories

Headband. Tambour beaded on black net with yellow seed beads and black and pale (faded) green bugle beads. The center is adorned with molded glass beads hand painted to accentuate the designs. Three pharaohs in cobalt blue glass and the larger cameo of an Egyptian woman's profile, also in glass, represent the fashion trends. Following the discovery of King Tutankhamun's tomb, in 1922, a market for Egyptian motif in fashion accessories emerged. c.1920s. $100-$150

Detail.

Choker or headband. Seed beads, gold steel beads, and black bugles; all bead work is attached to velvet ribbon. c. early 1900s. $75-100. *Courtesy Sandy Osborn*

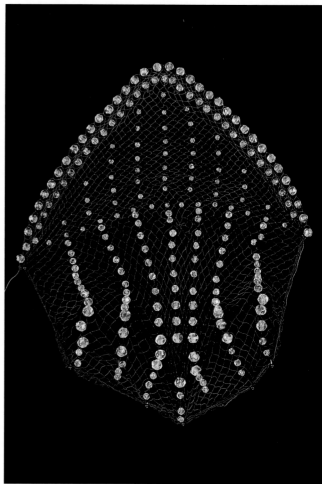

Top left: Netting crocheted with faceted crystal beads, creating a transparent hair covering, c.1920s. $75-100. *Courtesy Margaret Thorpe*

Top right: Back view.

Bottom left: Straw boater hat with black velvet appliqués on three sides; hand beaded with jet bugles, c.1930s. $100-150. *Courtesy Anna Greenfield*

Bottom right: Detail.

Top left: Black beaded beret, trimmed in rhinestones with a rhinestone clasp and silk tassel. c.1940s. $100-125. *Courtesy Christine Attenson, Attenson's Antiques and Books*

Top right: Jeweled velvet cocktail hat in a swirl pattern, beaded with pearls, silver-lined glass beads, and pearl and plastic drops. The front bow is created with amber rhinestones. c.1950s. $100-125. *Courtesy Mitchell S. Attenson, Attenson's Antiques and Books*

Bottom left: White straw embroidered with white china beads, trimmed with velvet flowers, c.1940s. $75-100. *Courtesy Linda Bowman, Legacy Antiques and Vintage Clothing*

Bottom right: Mourning hat in black velvet and lace; hand beaded in jet bugles on braided appliqué, c.1885. $100-125. *Courtesy Ursuline College Historic Costume Study Collection*

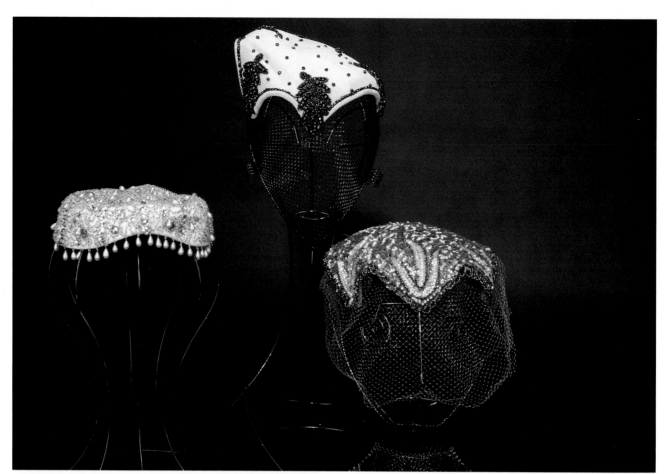

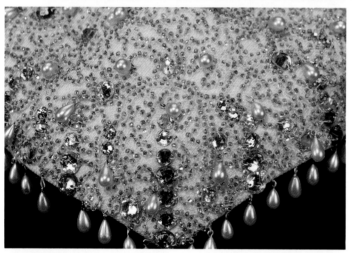

Left: black velvet pill box beaded in rows using pearls; glass beads with net tie.
Right: cream lace beaded with pearls and silver-lined bugles with a net. The Halle Bros. Co. label.
c.1950s. $50-75 each. *Courtesy Ursuline College Historic Costume Study Collection*

Left: black satin with black cording bordering crescent shapes that are jet beaded. Schiaparelli designer, made in Paris. c. 1950s. $100-125. *Courtesy Charlotte Michell Trenkamp.*
Right: brown satin crown with tambour stitched, iridescent sequins, glass amber cut beads in a "vermicelli" pattern. c.1960s. $50-75. *Courtesy Ursuline College Historic Costume Study Collection*

Opposite page:
Top:
Left: white felt pill box encrusted with silver-lined bugles in vermicelli pattern, scattered with round pearls and rhinestones and edged with drop pearls.
Top: cream velvet with leaf appliqués, beaded with black iridescent cut glass beads.
Right: brown felt covered with tambour stitched iridescent sequins and black cut beads.
c. late 1950s and early 1960s. $50-75 each. *Courtesy Anna Greenfield*

Bottom three: Details.

Above: Detail.

Top left: Left: brown felt, turned brim with brown satin cording bordered with gold glass cut beads in curl pattern. The A. Polsky Co., Akron, Ohio label.
Right: brown pillbox with brown satin thread embroidery bordered with gold metal seed beads. Mr. D. designer.
c.1960s. $75-100 each. *Courtesy Ursuline College Historic Costume Study Collection*

Bottom left: Left: blue velvet embroidered with white rice pearls, silver-lined beads, pearl leaves, and rhinestone studs. Glass discs and beads create floral center. c.1940s.
Right: navy blue velvet, embroidered with iridescent cut beads, bugles, and blue pearls; rhinestone studs with a net veil. c.1950s.
$50-75 each. *Courtesy Linda Bowman, Legacy Antiques and Vintage Clothing*

Black crocheted angora with embroidery in red angora yarn; hand beaded with faceted plastic black and red beads. Christian Dior designer. c.1960s. $100-125. *Courtesy Barbara Primeau Bachman*

Pale pink crown with wide turned up brim covered with netting; tambour stitched, embroidered with black sequins and jet beading in a leaf pattern, c.1960s. $100-125. *Courtesy Anna Greenfield*

Top left: Matching baseball cap and ballerina slippers embroidered with multi-colored sequins held fast with clear crystal seed beads, c.1990s. $75-100 set. *Courtesy Saiede Tajbakhsh Baghery*
Top right: Black baseball cap embroidered with star pattern in multi-colored sequins held fast with crystal seed beads, c.1995. $40-60.
Bottom: Headbands. Tambour stitched, with colored seed beads. Made in China. c.1990s. $25-35 each. *Courtesy Lillian Birnbaum Horvat*

Top left: Necklaces, loom woven in pink, silver, and clear crystal seed beads, c.1920s. $100-125 each. *Courtesy Shirley Friedland*

Bottom left: Necklaces of off-loom beading in a lattice pattern using gold-lined, garnet, and iridescent seed beads, c.1930s. $50 each. *Courtesy Shirley Friedland*

Top right: Necklace, loom woven in black and gold seed beads, creating a chevron pattern and Arabic script symbol, c. early 1900s. $100-125. *Courtesy Saiede Tajbakhsh Baghery*

Botom right: Fringe collars or necklaces. Black and gold, and lavender with lilac bugle bead fringe strung from loom woven band. c.1960s. $50-75 each.

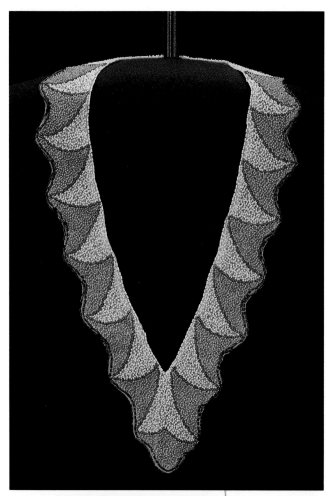

Above: Beaded collar. Dark green and gold beads with silver gray sequins on silk braid. c.1890s. $75-100. *Courtesy Pamela Nickel Wurster for Irene Hoffman*

Top left: Neckline trim, hand tambour beaded on beige crepe, in cream and tan seed beads with wine colored and white bugles, c.1930s. $75-100.

Bottom left: Detail.

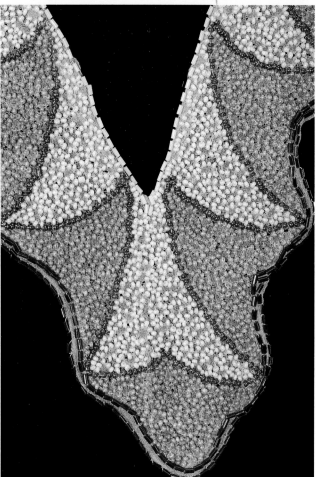

160

Top left: Red off-loom woven beaded collar in "lattice" pattern using opaque seed beads, c.1930s. $100-125.

Center: Red and gun metal gray off-loom woven beaded triangular collar in "lattice" design using opaque seed beads, c.1920s. $125-150.

Top right: Collar. Embroidered with black cut and bugle beads on grosgrain backing with matching Schiaparelli inspired bow earrings. c.1980s. $125-150 set. *Courtesy Mari Stanek Hageman*

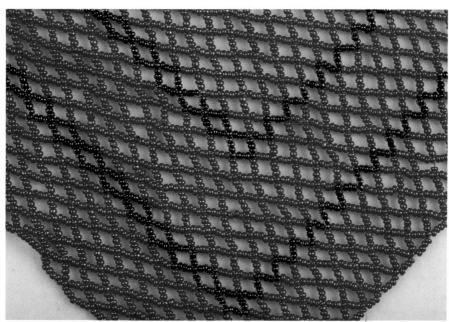

Detail.

Collar. White bead embroidery on silk using rice shaped pearls in a floral pattern around rhinestones with silver-lined seed and bugle beads. c.1930s. $75-100. *Courtesy Anna Greenfield*

Collar. Hand embroidered with white pearls on satin backing in even rows of 3mm pearls. The edge is trimmed in a looped fringe of pearls and a small fringe center. c.1950s. $50-75. *Courtesy Anna Greenfield*

Collar. White pearls that vary in size from 2mm to 6mm embroidered in even rows on silk. c.1950s. $70-80. *Courtesy Anna Greenfield*

Collar. White beads with jewel encrusted embroidery on satin with crystal beads creating a triangular pattern that is dotted with pearls and rhinestones. The fringe is created with crystal seed beads, pearls, and crystal faceted tear drops. c.1960s. $100-125. *Courtesy Linda Bowman, Legacy Antiques and Vintage Clothing*

Collar or shawl. Tambour stitched on white net with pearls and pale gold sequins. A 5-inch fringe is made of pearls. Sharmark, Las Vegas label. c.1980s. $150-200. *Courtesy Emma Lincoln*

Detail.

Top left: Collar or shawl. Pearls woven off-loom in eight bead and twelve bead patterns to create a lace netting and fringe. c.1960s. $125-150. *Courtesy Virginia Cangelosi Folisi*

Top right: Collar or shawl. Tambour stitched on black netting in sequins and gold-lined seed beads, c.1980s. $40-60.

Bottom: Red silk taffeta hip drape that is bead embroidered with charcoal bugle and cut beads in an Art Deco inspired design with looped fringe, c.1930s. $100-125. *Courtesy Anna Greenfield*

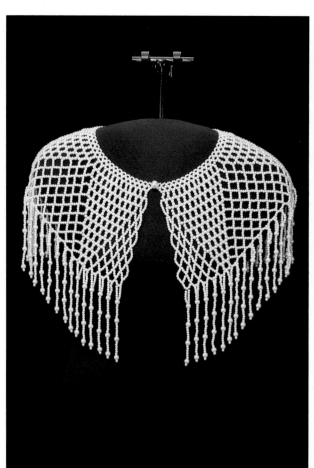

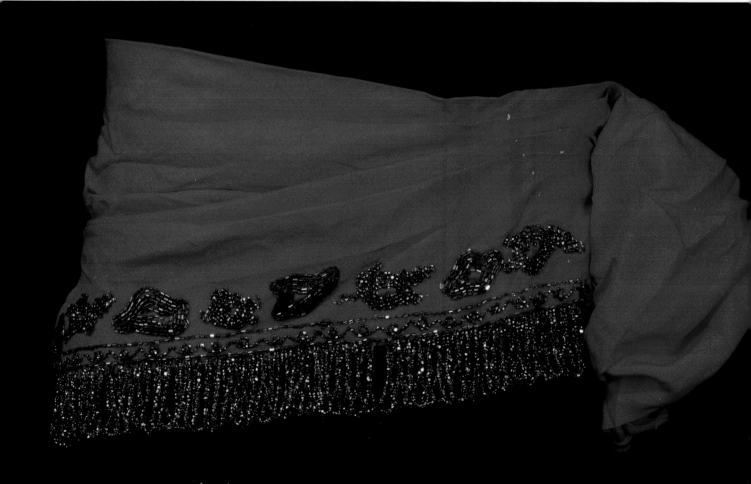

Top left: Pink and white braided belt of luster finish seed beads, with silk cording and tassel, c.1960s. $40-60. *Courtesy Shirley Friedland*

Bottom left: White beaded rope belt using seed beads braided around a cotton cord, c.1960s. $50-60.

Top right: White linked open weave pearl vest, c.1960s. $100-125. *Courtesy Virginia Cangelosi Folisi*

Bottom right: Black and silver beaded cummerbund, hand embroidered in bugle beads; with black cord ties, c.1970s. $35-40. *Courtesy Shirley Friedland*

Top left: Black silk chiffon scarf, tambour beaded with silver cut beads. Made in India. c.1930s. $50-75. *Courtesy Donna Kaminsky*

Top right: Black beaded sheer rayon scarf, hand embroidered with black seed and bugle beads in circle and diamond pattern, c.1960s. $75-100. *Courtesy Donna Kaminsky*

Bottom left: Tan silk chiffon scarf with multi-colored bead fringe, c.1960s. $50-75.

Bottom center: Teal silk chiffon tambour stitched in silk thread and teal bugle beads, c.1970s. $50-75.

Bottom right: White crepe shawl embroidered with silver-lined bugle beads, clear crystal seed beads, silver sequins and sequin discs at the end of bead fringe, c.1950s. $100-125. *Courtesy Emma Lincoln*

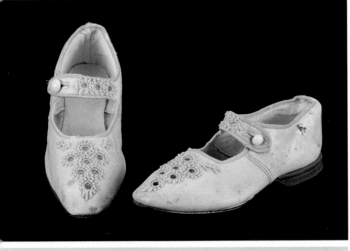

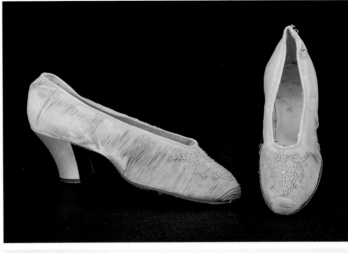

Top left: White child's leather shoes, hand embroidered in white luster finish seed beads, c.early 1900s. $50-75. *Courtesy Anne Katherine Brown, Friends Antiques*

Bottom left: Black satin evening pumps with beaded toe and tongue in cut jet beads, c.1927. $100-125. *Courtesy Charlotte Michell Trenkamp*

Top right: Pink silk pumps, hand embroidered in crystal seed beads, c.early 1900s. $75-100. *Courtesy Anne Katherine Brown, Friends Antiques*

Center right: Black satin evening pumps with steel micro beads, size 20 in red, blue, and black, c.1920s. $100-125. *Courtesy Linda Katz, Marc Goodman Antiques*

Bottom right: Detail.

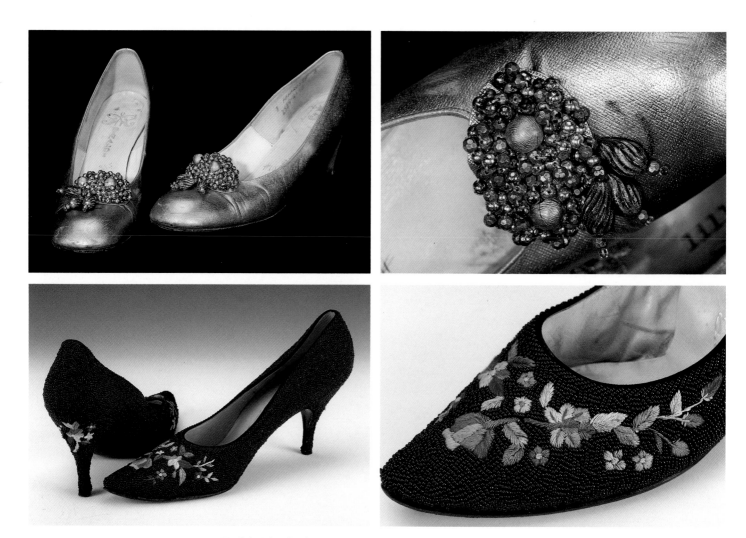

Top left: Silver leather pumps with silver and gold-colored bead clusters, c.1950s. $25-35. *Courtesy Anna Greenfield*

Top right: Detail.

Bottom left: Black beaded evening pumps. Cotton thread floral embroidery on toe and heel over black seed beads. c.1950s. $150-200.

Bottom right: Detail.

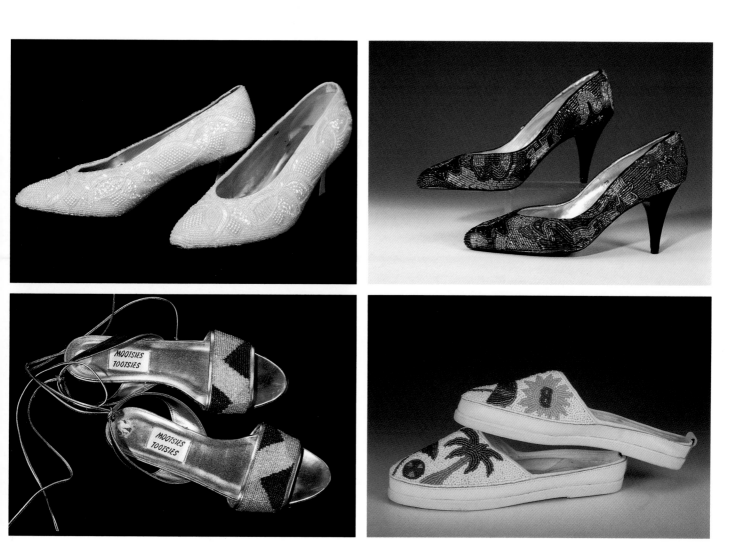

Top left: Ivory wedding pumps with pearls, bugles, sequins, and seed beads covering entire pump, c.1996. $50-75. *Courtesy Jill Korosec Dennis*

Top right: Multi-colored contemporary design high heels totally covered in seed beads, c.1980s. $75-100. *Courtesy Lillian Birnbaum Horvat*

Bottom left: Gold, black and silver sandals with seed beads forming a geometric motif; with a strap ankle wrap tie. Mootsies Tootsies. c.1980s. $30-40. *Courtesy Shirley Friedland*

Bottom right: Palm tree motif scuff sandals with multi-color seed beads, c.1998. $100-125. *Courtesy Bonnie's Goubaud*

The following Frankie & Baby and Caparros beaded flats, slippers, and tennis shoes are courtesy Donna Kaminsky.

White embroidered fabric accented with red, black seed, and bugle beads on the flowers and in the shape of a lady bug. Beverly Feldman, Frankie & Baby, style, "Lady Bug." c.1990s. $50-75.

Left: gold and white mesh embellished on the front by seed beads, "Jasmin."
Right: black suede embellished with a woven seed bead design.
c.1980s. $50-60 each.

Gold mesh shoes beaded with seed and bugle beads on heel and toe; also in tan linen and black suede. Beverly Feldman, Frankie & Baby, "Lisa." c.1980s. $50-60 each.

Gold, black, and bronze, beaded with seed and bugle beads. Beverly Feldman, Frankie & Baby, "Inspire." c.1990s. $50-75 each

White faille shoes embellished with seed beads, gold and silver braid with accent rhinestones. Caparros, "Alisha." c.1990s. $60-75.

Gold and black satin slipper, embellished with sequins, bugle beads, small and large seed beads with accent rhinestones. Beverly Feldman, Frankie & Baby, "Bali." c.1990s. $75-100 each.

Top: Black, gold, and silver color choice of slipper, embroidered in sequins, and seed and bugle beads. Caparros, "Leo." c.1990s. $70-90 each

Bottom left: White and in gold, embroidered with sequins, secured by individual seed beads. Caparros, "Attire." c.1990s. $50-60 each.

Bottom right: White opaque seed beads in all-over beading; heart design produced with gold and silver seed beads. Beverly Feldman, Frankie & Baby, "Crazy for You." c.1990s. $80-100.

Top left: Left: multi-color sequin and seed bead slippers.
Right: gold lamé slippers embellished with seed and bugle beads, sequins, and various crystal shaped accent beads.
Beverly Feldman, Frankie & Baby, "Holiday." c.1980s. $60-75 each.

Top right: Left: multi-colored slippers with sequins and seed beads.
Right: tan wooden seed beads, gold glass bugles, and larger accent wooden beads.
Caparros. c.1980s. $75-100 each.

Bottom left: Left: black and red sequin slippers with red bugle beads and red accent stones.
Right: iridescent purple seed beads with larger amber accent stones.
Valerie Stephens and Summit Hill. c.1990s. $75-100 each.

Bottom right: Green, bronze, blue, and khaki color choice of slippers; all are covered in seed beads in a plaid design. Caparros, "Tartan." c.1990s. $75-100 each.

Floral design featuring a dragon fly motif. Left: white, embroidered with pearls; design is formed by using sequins, and seed and bugle beads. Right: black pearls; design is formed by using sequins, and seed and bugle beads. c.1990s. $100-125 each.

Dragon fly motif. Red pearls with design formed with sequins, and seed and bugle beads. c.1990s. $100-125 each.

Detail.

174

Gold and silver; black, red, and gold; both pairs are covered with sequins, and seed and bugle beads. Caparros, "Aria." c.1990s. $70-90 each.

Black sequins accented with large colored rhinestones; tiny gold balls outline the accent beads. Caparros, "Haven." c.1990s. $70-90.

Left: teal iridescent seed beads cover these pull-on sneakers.
Right: red and blue slippers with embroidered seed beads.
Caparros, "Indian." c.1990s. $70-90 each.

Top left: Heart design in gold, bronze, and silver beads on black beaded tennis shoes. Beverly Feldman, Frankie & Baby, "Heart to Heart." c.1990s. $70-90.

Bottom: Assorted designs on tennis shoes embroidered with seed beads. "Beadz."c.1990s. $60-75 each.

Top right:
Moon and Stars design.
Left: black slipper.
Top: white with metallic shoe lace.
Right: black with metallic shoe lace.
Beverly Feldman, Frankie & Baby, "Moon time." c.1990s. $75-100 each.

Assorted designs on tennis shoes embroidered with seed beads and sequins. "Beadz." c.1990s. $60-75 each.

Pastel beaded tennis shoes and purse with a zig-zag pattern of seed beads, c.1990s. $50-75 each.

Pink and lavender slippers embellished with seed beads. "Sugar Traps." c.1990s. $50-75 each.

Left: turquoise slipper with sequins, and bugle and seed beads.
Right: fuchsia slipper with sequins, and bugle and seed beads.
Caparros, "Heat." c.1990s. $50-75 each.

Argyle designs in blue, red, and black color choice satin slippers embroidered with colored pearls, and bugle and seed beads. Beverly Feldman, Frankie & Baby, "Scotty." c.1990s. $70-90 each.

Black suede boots embroidered with sequins and gold balls, c.1990s. $100-150.

Bottom left: Black satin slippers embroidered with silk thread and accented with silver seed beads and sequins. Beverly Feldman, Frankie & Baby, "Lotus." c.1990s. $60-80 each.

Bottom right: Black satin embroidered slippers, accented with sequins, and seed and bugle beads. Beverly Feldman, Frankie & Baby, "Beauty." c.1990s. $50-75.

Top: American Indian inspired design.
Left: black embroidered velvet accented with silver seed beads and sequins.
Right: multi-colored design entirely created with seed beads.
Beverly Feldman, Frankie & Baby, "Kachina." c.1990s. $125-150 each.

Bottom: Detail.

Top: Multi-color design created with seed beads. Caparros, "Andrea." c.1990s. $100-125.

Bottom: Detail.

Multi-color black, silver, and gold; embellished with bugle beads and matching sequins. Beverly Feldman, Frankie & Baby, "Mambo." c.1990s. $75-100 each.

Detail.

Interlocking shapes in sequins outlined with seed beads. Caparros, "Chain." c.1990s. $75-100 each.

City scene design on black and navy satin slippers featuring the Eiffel Tower embellished with sequins, seed, bugle, and accent beads. Beverly Feldman, Frankie & Baby, "City." c.1990s. $100-125 each.

Blue and green canvas slippers with sequins, and seed and bugle beads. Beverly Feldman, Frankie & Baby. c.1990s. $75-100.

Multi-color slippers embellished with sequins and bugle beads. Beverly Feldman, Frankie & Baby, "Park Avenue." c.1990s. $100-125.

Top left: Multi-color slippers with sequins, and seed and bugle beads. Beverly Feldman, Frankie & Baby, "Chrysler." c.1990s. $75-100.

Bottom left: Black mesh shoes embroidered and embellished with large seed beads, sequins, and bugle and gold beads. Beverly Feldman, Frankie & Baby. c.1990s. $100-125.

Top right: Chinese inspired design in two color choices embroidered in sequins fastened by seed beads, gold balls, colored pearls, and rhinestones. Beverly Feldman, Frankie & Baby, style "Canton." c.1990s. $80-100 each.

Bottom right: Detail.

Opposite page:
Top: Geometric design in two color choices embellished with seed beads, sequins, bugle beads, and large accent beads. Beverly Feldman, Frankie & Baby, "Geo." c. 1990s. $100-125 each.

Bottom: Detail.

Top: Paisley design in three color choices embellished with sequins and seed, bugle, and large beads. Beverly Feldman, Frankie & Baby, "Taj." c.1990s. $100-125 each.

Bottom: Detail.

Top: Fruit design produced with seed beads, bugle beads, sequins, and various shaped accent beads. Beverly Feldman, Frankie & Baby, "Banana." c.1990s. $100-125.

Bottom: Detail.

187

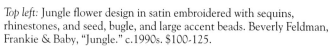

Top left: Jungle flower design in satin embroidered with sequins, rhinestones, and seed, bugle, and large accent beads. Beverly Feldman, Frankie & Baby, "Jungle." c.1990s. $100-125.

Bottom left: Detail.

Top right: American Indian mask motif hand embroidered in gold, charcoal, black crystal seed beads, and gold filled rocailles. Caparros, "Thunder." c.1990s. $125-150.

Bottom right: Detail.

Bibliography

Aghayan, Ray (Elizabeth Courtney Costumes, Studio City, California). Telephone interview, September, 1998.

Ashelford, Jane. *The Art of Dress, Clothes and Society 1500-1914*. London: National Trust Enterprises Limited, 1996. 320 pages. Numerous costumes, textiles, and historical information from the resources of Britain's National Trust with color illustrations.

Abiti in festa, l'ornamento e la sartoria Italiana. Galleria del Costume de Palazzo Pitti, Livorno, Italia: sillabe, 1996. 101 pages. Interviews with Italian designers and color illustrations of evening gowns from the Pitti Palace, (Florence, Italy) costume exhibit, March 30 to December 31, 1996. Italian text.

Atkins, Robin. "Process: Patricia Frantz." *Ornament*, Vol. 15, No. 3, (Spring 1992): 80-95. Description in text and color illustrations of lampwork beadmaking.

Barth, Georg J. *Native American Beadwork*. Stevens Point, Wisconsin: R. Schneider, Publishers, 1993. 219 pages. Brief History and complete instructions, with black and white illustrations of American Indian beadwork.

Batt, Paulette (Owner of Larchmere Antiques and Paulette's Vintage, Cleveland, Ohio. Expert in collecting and restoring vintage apparel; schooled in Belgium for needle work and bead emroidery.) Personal interview, February - November, 1998.

Berry, Sue (Western Reserve Historical Society, Chisholm Halle Costume Wing). Personal interview, September, 1998.

Blair, Joane. *Fashion Terminology*. New Jersey: Prentice Hall, 1992. 129 pages. Dictionary of fashion terminology with illustrations.

Brower, Mark (Associate Designer, Vera Wang, Ltd., New York). Telephone interview, July, 1998.

Broyles, Mary (President of Ornamental Resources, Inc., Idaho Springs, Colorado, a comprehensive order company for beads and related supplies.) Telephone interviews, July - August, 1998.

Buchholtz, Barbara B. "Stylish Collecting." *Chicago Tribune*, Section 15, p. 3, "Home," Sunday, October 26, 1997. Information on collecting vintage apparel.

Campbell-Harding, Valerie, and Pamela Watts. *Bead Embroidery*. Berkeley, California: LACIS, 1993. 128 pages. Instructions for bead embroidery on apparel with color illustrations. Video tape by the same title.

Campbell, Jean. *Beadwork Basics, Interweave Beadwork*. Loveland, Colorado: Interweave Press. 18 pages. Pamphlet with information and instructions for beadwork.

Christie, Manson, & Woods Ltd., 1997. *Madame Grès: The Archive Collection*, Thursday, 17 September, 1998. London, England. 88 pages. Christie's color illustrated catalog, 98 lots of a Madame Grès collection of gowns, with bid ranges and final sale prices.

Clabburn, Pamela. *Beadwork*. Aylesbury, Bucks, England: Shire Publications, 1980. 32 pages. Historical information with black and white illustrations.

Cleveland, Debra Regan. "French Designers and the Art Deco Movement." *Lady's Gallery* (Vol. I, Issue 4): 4-16. History and discussion of collecting Deco period apparel and accessories.

Coles, Janet, and Robert Budwig. *The Book of Beads: A Practical Guide to Beads and Jewelry Making*. New York: Simon & Schuster, 1990. 125 pages. Brief history, instructions, and comprehensive catalog of bead types in color.

_____. *Beads: An Exploration of Bead Traditions Around the World*. New York: Simon & Schuster Editions, 1997. 160 pages. Historical information and color illustrations.

Conner, Wendy Simpson. *The Best Little Beading Book*. California: Interstellar Publishing Company, 1995. 274 pages. Descriptions and comprehensive instructions with black and white photographs.

Costume. Cleveland, Ohio: The Western Reserve Historical Society, Chisholm Halle Costume Wing, 1986. 35 pages. Catalog of selections from the collection in chronological order, with descriptions of color illustrations.

Creighton, Janet. "Following the Trail of Glass." *American Archaeology*, (Winter 1997-98): 8- 12. Description of archaeological discoveries and history of 19th century glass trade beads in the Pacific Northwest.

Distinctive Details: Great Embellishment Techniques for Clothing. New York: Taunton Press, 1995. 95 pages. Collection of articles from *Threads* Magazine, with instructions and color illustrations of beading and sewing techniques.

Dolan, Maryanne. *Vintage Clothing 1880-1960: Identification & Value Guide*. Florence, Alabama: Books Americana, Inc., 1987. 219 pages. Historical information, design descriptions with black and white illustrations; price range included for most exhibits.

Druesedow, Jean L. *In Style: Celebrating Fifty Years of the Costume Institute*. Third printing, 1995; reprinted from *The Metropolitan Museum of Art Bulletin* (Fall 1987) by the Metropolitan Museum of Art, 1987. 64 pages. Catalog in color illustrations of historic costumes arranged in chronological order, with descriptions, and historic detail.

Dublin, Lois Sherr. *The History of Beads: From 30,000 B.C. to the Present*. New York: Harry N. Abrams, Incorporated, 1987. 364 pages. Comprehensive history of beads with full page color and black and white illustrations; time-line chart.

_____. *The History of Beads: From 30,000 B.C. to the Present, Concise Edition*. New York: Harry N. Abrams, Incorporated, 1987. 136 pages. Abridged edition of history of beads with color photographs.

Edwards, Joan. *The Bead Embroidered Dress*. Surrey, England: Bayford Books, 1985. 28 pages. Description of glass bead industry and history of bead embroidery in the nineteenth and twentieth centuries.

_____. *Bead Embroidery*. Berkeley, California: LACIS, 1992. 192 pages. Comprehensive history with research and sketches of beadwork in English museums.

Erikson, Joan Mowat. *The Universal Bead.* New York: W. W. Norton & Company, Inc., 1993. 191 pages. Cultural uses, history, and descriptions of materials.

Ewing, Elizabeth; revised and updated by Alice Mackrell. *History of Twentieth Century Fashion.* 3rd edition. New York: Quite Specific Media Group Ltd., 1992. 300 pages. Comprehensive history with black and white illustrations.

Finch, Karen, and Greta Putnam. *The Care and Preservation of Textiles.* Berkeley, California: LACIS Publications, 1991. 144 pages. Comprehensive guide to conservation of collection of costumes and textiles; with black and white illustrations.

Francis, Peter, Jr. *A Short Dictionary of Bead Terms and Types.* Peter Francis, 1979. 119 pages. Comprehensive, technical dictionary of beads and beading.

_____. *Beads of the World: A Collector's Guide With Price Reference.* Atglen, Pennsylvania: Schiffer Publishing, Ltd., 1994. 142 pages. History and descriptions of collectible beads, beaded items, and beadwork with color illustrations.

Fraser, Kennedy. *The Fashionable Mind: Reflections on Fashion, 1970-1981.* New York: Alfred A. Knopf, 1981. 270 pages. Commentary on the decade of fashion with black and white illustrations.

Friedman, Florence Dunn, ed. *Gifts of the Nile: Ancient Egyptian Faience.* Exhibit: The Cleveland Museum of Art and Museum of Art, Rhode Island School of Design. New York: Thames and Hudson, Inc., 1998. 288 pages. Comprehensive history with 483 illustrations in color and black and white.

Fukuyama, Yusai. *Tambour Work.* Berkeley, California: LACIS Publications, 1997. 120 pages. Comprehensive instructions with black and white illustrations.

Gimble, Frances. *After a Fashion: How to Reproduce, Restore, and Wear Vintage Styles.* San Francisco, California: Lavolta Press, 1993. 337 pages: Comprehensive instructions for identifying, collecting, reproducing, and restoring vintage apparel; lists of resources.

Giunca, Mary. "Fortuny of Venice," *Lady's Gallery* (Vol. III, Issue 2): 8-14. Color illustrations from the sales brochure, Arthur H. Lee & Sons, Inc.; with history and description.

Goldin, Eileen. *Beads and Sequins.* South Africa: National Book Printers, 1989. 32 pages: Instructions for bead embroidery with color illustrations.

Haertig, Evelyn. "Venice And Its History of Beads," *Lady's Gallery* (Vol. III, Issue 2): 16-20, 47-48. An edited chapter from *More Beautiful Purses* by Evelyn Haertig; with color illustrations.

Harris, Elizabeth. *A Bead Primer.* 6th Printing. Arizona: The Bead Museum, 1987. 30 pages. Pamphlet; descriptions and history of glass beads with black and white illustrations by Karen Lindquist.

Harris, Kristina. *Vintage Fashions for Women, 1920s-1940s, with Values.* Atglen, Pennsylvania: Schiffer Publishing, Ltd., 1996. 192 pages. Guide to collecting with color illustrations of modeled apparel.

_____. "Vintage Vine," *Lady''s Gallery* (Vol. I, Issue 2): 41. Response to question on collecting vintage apparel.

_____. "Vintage Vine," *Lady's Gallery* (Vol. II, Issue 1): 48. Response to question on price guides for vintage apparel.

Haug, Joanne. "Stepping Back in Time With the Bata Shoe Museum Collection." *Lady's Gallery* (Vol. I, Issue 4): 36-41. History, descriptions, and color illustrations of selections from the collection.

"Hold Onto Your Hats: The History and Meaning of Headwear in Canada," *Lady's Gallery* (Vol. IV, Issue 4): 44-45. Description of exhibit of vintage head wear at the Canadian Museum of Civilization, Hill, Quebec, 1997; with color illustrations and history.

Jacobs, Laura, and Victor Skrebneski, photographs. *The Art of Haute Couture.* New York: Abbeville Press, 1995. 176 pages. History and descriptions with color illustrations by a leading fashion photographer.

Jargstorf, Sibylle. *Glass Beads from Europe.* Atglen, Pennsylvania: Schiffer Publishing, Ltd., 1995. 192 pages. History of glass beads with color illustrations.

Jerde, Judith. *Encyclopedia of Textiles.* New York: Facts On File, Inc., 1992. 260 pages. Comprehensive descriptions with color and black and white illustrations.

Jick, Millicent. "Bead-Net Dress From Giza Tomb G7740z, Old Kingdom Dynasty IV, Reign of Khufu," *Ornament* Vol. 14 No. 1 (Autumn 1990): 50-53. Color illustrations including restored faience bead-net dress; with descriptions and historic documents.

Jones, Julia. *The Beading Book.* Berkeley, California: LACIS Publications, 1993. 142 pages. Brief history and comprehensive instructions for beadwork with color and black and white illustrations; includes guide to care and conservation.

Karklins, Karlis. *Glass Beads: The 19th Century Levin Catalogue and Venetian Bead Book and Guide to Description of Glass Beads.* Canada: Minister of Supply and Services, 1985. 123 pages. Technical catalog of a collection (by Moses Levin) of glass and stone beads dating 1851-1869.

Kennett, Frances. *The Collector's Book of Fashion.* New York: Crown Publishers, Inc., 1983. 256 pages. Comprehensive history of fashion from 1900 to 1970's with sections on accessories, men's and children's wear; with black and white illustrations and some color.

Klinkenborg, Verlyn. "Native American Beadwork: Fields of Colored Glass Define Traditional Tribal Patterns," *Architectural Digest* (June 1992): 132-137, 191. Color illustrations and descriptions of beadwork of the American Plains Indians.

Kock, Jan, and Torgen Sode. *Glass, Glassbeads and Glassmakers in Northern India.* Denmark: THOT print. 30 pages. Personal account of exploring the glass producing centers of Northern India with color illustrations.

Laver, James. *Costume and Fashion: A Concise History.* Reprinted 1986. New York: Thames and Hudson, 1969-1982. 288 pages. Comprehensive history with black and white (some color) illustrations.

Leaf-Hund, Doreen (Designer and nationally recognized expert in restoration of vintage apparel; specialized in couture embroidery with beads; continues as a student of the Lesage School, Paris, France.) Personal interviews, January - November, 1998.

Liese, Gabrielle (Director and founder of The Bead Museum, Prescott, Arizona.) Personal interview, January, 1998.

Liu, Robert K. *A Universal Aesthetic: Collectible Beads.* California: Ornament, Inc., 1995. 256 pages. Comprehensive descriptions and historic reference with color illustrations.

McConathy, Dale, with Diana Vreeland. *Hollywood Costume.* New York: Harry N. Abrams, Inc., 1976. 317 pages. A retrospective discussion of the influence of Hollywood designers on dress and fashion; with color illustrations.

Mailand, Harold F. *Considerations for the Care of Textiles and Costumes: A Handbook for the Non-Specialist.* Indiana: The Indianapolis Museum of Art, 1980. 23 pages. Basics of techniques and practices of conserving collections of costumes and textiles.

Mandelbaum, Howard, and Eric Myers. *Forties Screen Style: A Celebration of High Pastiche in Hollywood.* New York: St. Martin's Press, 1989. 209 pages. History with black and white illustrations.

Martin, Richard. *Gianni Versace.* New York: Harry N. Abrams, Inc., 1997. 191 pages. Published in conjunction with exhibit by the same name at The Metropolitan Museum of Art, from December 11, 1997, through March 22, 1998; color illustrations with descriptions.

Martin, Richard, and Harold Koda. *Haute Couture.* New York: Harry N. Abrams, Inc., 1995. 118 pages. Published in conjunction with exhibit by the same name at The Metropolitan Museum of Art, from December 7, 1995, through March 24, 1996; a brief history and color illustrations with descriptions.

McConathy, Dale, with Diana Vreeland. *Hollywood Costume*. New York: Harry N. Abrams, Inc., 1976. 317 pages. A retrospective discussion of the influence of Hollywood designers on dress and fashion; with color illustrations.

Mellish, Susan. "Romancing the Past with Hats," *Lady's Gallery* (Vol. 1, Issue 5): 4-18. Color illustrations of vintage hats from private and museum collections; with history and descriptions.

Metzger-Krahe, Frauk. *Glasperlenarbeiten im 19. Jahrhundert*. Nienburg, Germany: 1980. 72 pages. Catalog published in conjunction with an exhibit by the same name (Glass Beadwork of the Nineteenth Century) at the Nienburg Museum; descriptions of bead embroidery and black and white illustrations; German text.

Milbank, Caroline Rennolds. *Couture: The Great Designers*. New York: Stewart, Tabori & Chang, Inc., 1985. 431 pages. Comprehensive, color illustrated history of the couture design work from the 1850s.

Milbank, Caroline Rennolds. *New York Fashion: The Evolution of American Style*. New York: Harry N. Abrams, Inc., 1989. 304 pages. Comprehensive history of American fashion from the 19th century to present, with black and white illustrations.

Moffhet, Laura (Director of Public Relations, Mary McFadden Couture, New York). Telephone interview, July 1998.

O'Keeffe, Linda. *Shoes: A Celebration of Pumps, Sandals, Slippers and More*. New York: Workman Publishing, 1996. 507 pages. Comprehensive selection of color illustrations, with history and descriptions.

Rothstein, Natalie. *Four Hundred Years of Fashion*. Reprint, 1992. London: Victoria and Albert Museum, 1984. 176 pages. Description, history, and color illustrations of costume and accessory collections.

Sears, Roebuck Catalogue, 1908: A Treasured Replica from the Archives of History. Joseph J. Schroeder, ed. Illinois: DBI Books, Inc., 1971. 1184 pages. Republication of the original catalogue.

Shaeffer, Claire B. *Couture Sewing Techniques*. Reprint. Connecticut: The Taunton Press, Inc., 1993. 217 pages. Discusses haute couture fashion, history, collecting, and a comprehensive instruction of fine sewing; with black and white and some color illustrations.

Smith, Pamela. *Vintage Fashion and Fabrics: Instant Expert*. New York: Alliance Publishing, Inc., 1995. 153 pages. Pocket size guide to collecting vintage apparel, with black and white illustrations and resources guide.

Sotheby's. *Paris A La Mode . . . Haute Couture*, October 29, 1997. New York: Sotheby's. 130 pages. Catalog, inaugural sale of Sotheby's New York Fashion Department of Parisian "haute couture," with color illustrations, descriptions, bid ranges, and sale prices.

Steinberg, Donna. "Buckskin Dresses." *Threads Magazine* (Aug-Sept 1989): 62-65. Description of traditional dress of Southern Plains Indian women, with instruction and color illustrations.

Steiner, Christopher. "Symbols of Tradition: West African Trade Beads," *Ornament*, Vol. 14, No. 1 (Autumn 1990): 32-34. Discussion of the popularity of the trade beads found in Africa that are now being sold to collectors throughout the world.

Taylor, Lou. *Mourning Dress: A Costume and Social History*. London: George Allen and Unwin, 1983. 327 pages. Comprehensive history of the traditions of mourning; with black and white illustrations.

Thompson, Angela. *Embroidery With Beads*. London, England: B. T. Batsford, 1987. 120 pages. History and comprehensive instructions in techniques; with color and black and white illustrations.

Tsosie, Michael Phillip. "Historic Mohave Bead Collars." *American Indian Art Magazine*, Vol. 18, No. 1 (Winter 1992): 36-49. History of 19th century and contemporary beadwork; with descriptions from documents with black and white illustrations.

Van der Sleen, W. G. N. *A Handbook on Beads*. New York: George Shumway, 1966. 128 pages. Technical study of trade beads with descriptions and names compared in six languages.

Versace, Gianni. *Vanitas: Designs*. New York: Abbeville Press, 1994. 272 pages. Color illustrations of Versace's designs and drawings.

Wayne, George. "Puttin' On the Glitz: Bob Mackie's Showgirl Chic." *Music Choice*. June 1998: 20-21. An interview with Bob Mackie about his designs for stars, Cher, and Ann-Margaret; with color photos.

White, Mary. *How To Do Bead Work*. New York: Doubleday, Page and Company, 1904. Reprint. New York: Dover Publications, Inc., 1972. 142 pages. Instructions for beadwork projects, with black and white illustrations.

White, Palmer. *Haute Couture Embroidery: The Art of Lesage*. Originally published in France: Editions du Chene, 1987. Republished, Berkeley, California: LACIS Publications, 1994. 171 pages. Comprehensive history of Lesage family couture embroidery and embellishment; with color illustrations.

Wilcox, Claire, and Valerie Mendes. *Modern Fashion in Detail*. New York: The Overlook Press, 1991. 143 pages. Collection of couture apparel featuring design details; color illustrations.

William Doyle Galleries. *Couture and Textiles*, December 5, 1996. 83 pages. Catalog of 764 lots, with descriptions, black and white illustrations, bid ranges, and sale prices.

_____. *Couture and Textiles*, June 11 and 12, 1997. 86 pages. Catalog of 765 lots, with descriptions, black and white illustrations, bid ranges, and sale prices.

_____. *Couture and Textiles*, November 12, 1997. 107 pages. Catalog of 792 lots, with descriptions, black and white illustrations, bid ranges, and sale prices.

_____. *Couture and Textiles*, May 6, 1998. 102 pages. Catalog of 587 lots, with descriptions, black and white illustrations, bid ranges, and sale prices.

Wilson, Elizabeth, and Lou Taylor. *Through the Looking Glass: A History of Dress From 1860 to The Present Day*. London, England: BBC Books, 1989. 240 pages. Comprehensive text that accompanies the BBC TV series by the same name; black and white illustrations.

Index

Note: in most cases, unless designer names are followed by a company designation in the text or caption, i.e., Co., Ltd., Inc., they are alphabetized as individuals, e.g., Mackie, Bob.